Titanic:
A Legal Perspective
by E. Gordon Mooneyhan

Titanic: A Legal Perspective
Copyright 2023
E. Gordon Mooneyhan

Published by
Sea Island Publishing
1413 Hwy 17 S., #377
Surfside Beach, SC 29575

ISBN-13: 9798396642300

Other books by E. Gordon Mooneyhan

The Southern Railway Dining Car Cookbook

The Atlantic Coast Line Railroad Dining Car Cookbook

The Seaboard Air Line Railroad Dining Car Cookbook

My Journey Through Congestive Heart Failure

Public Relations for the Volunteer Public Information Officer

Dedication

The dedication of this book is to the memory of the 1,496 people who died in the sinking of the RMS Titanic on April 15, 1912. May they never be forgotten.

CONTENTS

Acknowledgements 11

Introduction 13

Ch 1 Negligence, Prudence, and Collusion 21

Ch 2 The Weather on April 14, 1912 27

Ch 3 Cold Weather Mirage 31

Ch 4 The Wireless Operators 35

Ch 5 Navigation 55

Ch 6 The Importance of Time to Celestial Navigation 69

Ch 7 Lifeboats 83

Ch 8 J. Bruce Ismay 89

Ch 9 Captains Smith, Rostron, and Lord 103

Ch 10 A Series of Mistakes 113

Ch 11 A Case of Collusion 129

Ch 12 The Marconi Company Practices 137

Ch 13 What if… 147

Ch 14 Conclusions 155

Bibliography 163

Acknowledgments

Several people need to be thanked as their contributions have helped improve the quality of this work. I am grateful to Tom Winslow of Winslow Law for agreeing to read the manuscript and for helping me understand the legal nuances I had to deal with in writing this book. This project differed significantly from Principles of Business Law in college and was much more challenging.

More years ago, than I care to remember, I took a Navigation and Seamanship course at the University of South Carolina, Conway Campus, now Coastal Carolina University (CCU). I had a basic understanding of how to use a sextant and how to do celestial navigation, but time and not using that knowledge had caused me to forget virtually all of what I had learned. Thankfully, the internet allowed me to relearn the basics and convey that knowledge in this book.

There were a great many users on the Encyclopedia-Titanica.org website who were more than willing to offer their opinions and answers to the many questions I posted. If I tried to list everyone, I would likely forget some, and I am thankful to them all.

Likewise, Nancy Goettel, my computer science instructor from CCU, was willing to answer my questions regarding computer simulations, mainly concerning whether or not simulations can be "manipulated" to yield a predetermined result.

The internet and email are genuinely beautiful inventions. Not only have they made the world smaller and researching a topic easier, but email especially has made it possible to communicate with people you would otherwise not have a chance to meet. Because of the internet and email, I have become acquaintances with relatives of some people from that historic night, and I've had a new understanding of the survivors.

Last but certainly not least, Samuel Pence, aka Historic Travels on YouTube, rekindled my interest in that grand lady of the sea, the Royal Mail Ship Titanic. Sam, thank you so much for inspiring me to write this book.

Introduction

Perhaps no maritime accident has had more written about it than the sinking of the R.M.S. Titanic on April 14-15, 1912. Of course, it's not the worst marine accident in terms of lives lost; that dubious honor belonged to the MV Wilhelm Gustloff when more than 9,000 souls perished near the end of WW2. But the Titanic intrigues people. It was the ship of dreams; at the time, it was the largest movable object created by man, displacing 52,310 tons of water. The owners (White Star Line) never claimed that she was "unsinkable" The shipbuilder (Harland and Wolff) said, "It is as unsinkable as we know how to make it." White Star Line brochures called her "practically unsinkable." But the press condensed that to "unsinkable," and White Star Line never denied it. Contributing to the "unsinkable" mythos was the collision of Titanic's sister ship, the R.M.S. Olympic, with the H.M.S. Hawke. The Hawke had a reinforced bow designed to allow the ship to ram enemy vessels to disable or even sink them. The Olympic escaped with relatively minor damage, thus contributing to the unsinkable legend.

The wealthiest man in the world, John Jacob Astor, IV, was on board. So were former Congressman Isidor and his wife Ida Straus, co-owners of Macy's, as were hundreds of immigrants hoping to make a new life in the New World. All told, 2,208 passengers and crew were on board, of which 712 survived.

The weight of a ship will determine how much water it displaces. Ships float because they displace more water than they weigh. For example, the Titanic displaced 52,310 tons of water. As long as a ship displaced more water than she weighed, she would float. However, once the weight of the water on board exceeded the weight it displaced, she would sink, and the laws of physics could not change that fact.

The U.S. Senate and British Board of Trade Inquiries held Captain Edward J. Smith of the Titanic responsible for the tragedy. It was a foregone conclusion; the ship's captain is ultimately responsible for the safety of the vessel, crew, and passengers. And it's always easier to blame someone who can't be there to defend themselves; ergo, the blame fell on Captain Edward John Smith.

I have used multiple primary sources for the research for this book. First, I obtained copies of the U.S. Senate and British Board of Trade Inquiries from www.titanicinquiry.org. Second, Walter Lord's (no relation to Captain Stanley Lord) "*A Night to Remember*" is perhaps the most accurate firsthand account of the sinking. Between 60 and 70 survivors of the disaster helped him compile the timeline of the events of that horrific night, published 43 years after the accident. Third, Tad Fitch, Ken Layton, et al. wrote "*On A Sea of Glass,*" which is probably the best reference book on the Titanic, and gives virtually an hour-by-hour account of the entire voyage, switching to minute-by-minute on the night of April 14-15, once the collision occurred. Fourth, Titanic Calling Wireless Communications During the Great Disaster is a compilation of almost all the Marconi wireless messages during the Titanic's sinking. The Marconi Company originally compiled this collection for the British Board of Trade Inquiry into the sinking. "*Titanic the Tragic Story of the Ill-Fated Ocean Liner*" by Rupert Matthews is a good, overall accounting of the Titanic's too-brief history. Lastly, www.encyclopedia-titanica.org is an online source for everything related to the Titanic. However, as with all online sources, I did try to find additional offline evidence to substantiate the online claims. Another book I discovered late in my writing, "*Titanic Signals of Disaster*" by John Booth and Sean Coughlan, is another collection of Marconigrams from that fateful night. Lastly, I met a cousin of J. Bruce Ismay's online (who also happens to be a Lodge Brother), and he suggested that I read the book *"The Ismay Line, "* a history of the White Star Line. In the process, I also discovered more about J. Bruce Ismay. There was much more to the man than the image presented by William Randolph Hearst's yellow journalists.

Since the sinking, we have learned much about weather conditions, radio signal propagation, where the Titanic was vs. where they thought they were, etc. That knowledge must be considered in any investigation. As a result, this book and its conclusions will likely be a hybrid of what happened at the time, tempered by what we know, or can reasonably hypothesize about conditions as they existed on the night of April 14-15, 1912, in the North Atlantic, roughly 400 miles southeast of Cape Race, Newfoundland.

A note about navigation: this book will use the navigational conventions in use in 1912. Locations are given in degrees, minutes, and seconds, i.e., 41° 43' 57.0" N; 49° 56' 48.8" W (read as 41 degrees, 43 minutes, 57.0 seconds North; 49 degrees, 56 minutes, 48.8 seconds West). This location is the generally accepted position of the *Titanic* wreck, based on the site of the boilers when Dr. Robert Ballard discovered the wreck in 1985, rather than decimal degrees (41.7325°N, 49.9469°W), which the advent of G.P.S. has made possible. There are 60 seconds in one minute and 60 minutes in one degree, but in terms of navigation, they do not relate to time. Therefore, when using the wreck location for calculations in this book, I have rounded the location to 41° 44' N, 49° 57' W. The rounding introduces an error of less than one-quarter mile into the calculations and keeps the figures in coordination with the other figures at the time.

The Titanic's discovery by Dr. Robert Ballard proved that the Titanic was not where she was reporting herself to be that night. It also confirmed that the Titanic broke into three pieces as she sank. However, unless otherwise noted, the positions used will be those reported by the Titanic, some 13 miles west of the actual wreck site.

In calculating the distances between various locations, I used the National Hurricane Center's site, Latitude/Longitude Distance Calculator (noaa.gov), which calculates distances using the great Circle Routes, giving the shortest distance between two points on the surface of a sphere.

I have endeavored to remain faithful to the facts and evidence presented at the hearings and written accounts of the wreck as demonstrated by the survivors. At the same time, I choose to look at the evidence presented with an open mind. And there appears to be omitted evidence, which needs to be considered.

Perhaps the greatest collection of evidence not presented at either hearing came in 1935, when Second Officer Charles Lightoller wrote his autobiography, *"Titanic and Other Ships,"* revealing information that would have made a significant difference in the 1912 inquiries. So why was that information not disclosed in 1912? Could it be that Lightoller's information in his book wasn't accurate, and he knew it? Or could it be that his mind was playing tricks on him when he was writing his autobiography?

There are two different Marconi's that are discussed in this work. In most cases, whether I am talking about Marconi, the man, or the company should be obvious. However, to avoid misunderstandings, when referring to Marconi the man, I will say either Mr. Marconi or Guglielmo Marconi. Any references to just "Marconi" refer to the company.

What has become evident in researching the history of this great ship is that things aren't always what they seem to be. As I told a fellow contributor on the Encyclopedia-Titanica website, the Titanic's story presents an endless series of rabbit warrens to explore, or, if I may be so bold as to paraphrase Sir

Winston Churchill, the Titanic is a riddle wrapped in a mystery, inside an enigma. Ultimately though, there comes a time when you need to stop researching and start writing.

CHAPTER 1

Negligence, Prudence, and Collusion

Although this is not a legal book, it helps to understand several legal terms as we look at the events surrounding the sinking of the RMS Titanic on April 14-15, 1912. Those terms are negligence, prudence, and collusion.

Negligence describes a situation in which a person acts careless (or negligent), resulting in someone else getting hurt or property getting damaged. Negligence can often be challenging to define because it involves a legal analysis of the elements of negligence as they relate to the facts of a particular case.[1] The plaintiff must show that the defendant was at fault for the injury or damage. Additionally, there are five parts to proving negligence:

1. Duty—duty arises when the law recognizes that a relationship exists between the plaintiff and the defendant, with the defendant required to act in a certain manner, often with a standard of care, towards the plaintiff.
2. Breach of Duty—a breach of duty arises when the defendant fails to recognize reasonable care in fulfilling the duty.
3. Cause in Fact—the plaintiff must prove that the defendant's actions were the cause of the

injury. This is often called the "but-for" causation, meaning that but for the defendant's actions, the plaintiff's injury would not have occurred.

4. Proximate Cause—Proximate cause relates to the scope of a defendant's responsibility in a negligence case. Defendants in negligence cases are only responsible for those harms that the defendant could have foreseen through their actions.

5. Damages—in a negligence case, the plaintiff must prove a legally recognized harm, usually in physical injury or death. It's not enough that the defendant failed to exercise reasonable care; the defendant's actions must result in actual damages to a person to whom the defendant owed a duty of care.

In addition to the above, the law of negligence has created the "reasonable person," which is, to a degree, a certain amount of legal fiction. It's based on how an ideal person would act in those identical circumstances. In effect, it is exercising the standard of care that is required in any given situation. It makes no allowance for "human error."

Several other factors are potentially involved in negligence claims; Contributory Negligence, Comparative Negligence, and Assumption of Risk.

Except for the Assumption of Risk, based on the events of the sinking of the Titanic, those factors don't enter into this discussion.

The second factor that night involved prudence which is the skillful or wise management of affairs; attentiveness to possible hazards, or caution or circumspection as to danger or risk; in other words, knowing the potential threats and drawing on your knowledge and experience to mitigate or avoid them. A person can act imprudently yet not suffer punishment because there are no consequences. For example, I could be driving my car on a busy highway, going 40 MPH over the speed limit, and swerving in and out of traffic. By some miracle, no accidents occur, and no one is injured. Yes, I was speeding, and my actions were imprudent, but no one has a cause of action against me because no one suffered any harm or damage (assuming that I managed to avoid encountering law enforcement).

Imprudence differs from negligence in that there are no damages for imprudence to occur, whereas negligence requires damage to have happened.

Lastly, collusion is "a secret understanding between two or more persons to gain something illegally, to defraud another of their rights, or to appear as adversaries though in agreement.[2]" There may have been collusion between Titanic's Second Officer Charles Lightoller and Fourth Officer Joseph Boxhall (or possibly all of Titanic's four senior surviving officers) during their testimony before the British Board of Trade Inquiry. We will look at this in greater detail in Chapter 11, A Case of Collusion.

ENDNOTES

[1] www.findlaw.com

[2] Dictionary.com. Last accessed October 30, 2022.

CHAPTER 2

The Weather on April 14, 1912

Many factors contributed to the sinking of the Titanic, some natural, some human. Let's start with the weather on April 14.

It was clear and relatively mild for most of the day on April 14, with temperatures hovering around 55°; all things considered, it was a pleasant day for mid-April in the middle of the North Atlantic Ocean, primarily due to the ship being in the Gulf Stream. However, between 7:00 and 7:30 p.m., the temperature began to plummet, dropping about 25° in a half hour. Two things account for this. First, the ship had crossed from the relatively warm Gulf Stream into the frigid Labrador Current; second, it had also passed through a cold front. Estimates are that the water temperature around the Titanic was about 28-29° Fahrenheit. While freshwater freezes at 32° F, seawater freezes at an average of 28.4° F because of the salt content, so it is fair to say the water was just about freezing. A healthy person plunged into water at that temperature could expect to survive for a maximum of 45 minutes, and most likely far less because of the shock that the body would endure getting plunged into water that was that cold.

There was concern about the temperature of the seawater affecting the freshwater stored on the Titanic. Second Officer Lightoller told Robert Hichens, one of the Titanic's quartermasters, to monitor the freshwater closely since it would indicate ice in the area if it began to ice.[1]

The sea was unusually calm that evening; Colonel Archibald Gracie wrote, "The sea was like glass, so smooth that the stars were clearly reflected."[2] There were reports that the sea was so calm it was challenging to determine where the horizon was. As a result, it would be difficult to do accurate celestial navigation after the end of nautical twilight, which could have contributed to the Titanic's crew determining an erroneous position and being 13 miles West of where they were.

One of the aids that mariners use to spot icebergs is the breaking of waves against the ice where it meets the water. This condition did not exist that night; the sea was dead calm. Although a perfectly calm ocean does not always indicate ice being nearby, the number of ice reports received during the journey should have raised the crew's situational awareness that the Titanic was closer to field ice than any of the ship's officers suspected. Unfortunately, the lookouts in the crow's nest did not have binoculars. The lack of wave action against the base of the iceberg would have rendered binoculars useless. Also, binoculars reduce the person's field of view, making it more difficult to spot the silhouette of the iceberg against the background stars, especially on a moonless night.

An average person's field of vision is approximately 95° to the side, 75° down, 60° inwards, and 60° above.[3] That's why television and movies are in the landscape format. Binoculars (and telescopes) work by narrowing the field of vision, which causes a smaller area to fill the same area, thereby magnifying the image.

ENDNOTES

[1] https://en.wikipedia.org/wiki/Charles_Lightoller Last accessed on September 23, 2022

[2] Gracie, Archibald, The Truth About the Titanic, M. Kennerley, New York (1913)

[3] Dr. Michael Questell, healthtap.com, last accessed on September 23, 2022

CHAPTER 3

Cold Weather Mirage

Everyone is familiar with mirages in hot weather—the shimmering "water" you see on the highway but never reach is perhaps the best example. What is not so widely known is that mirages can also occur in cold weather.

Light rays refract or bend over longer distances in cold air, causing objects below the horizon to appear above the horizon or a superior mirage. For example, figure 1 shows a superior mirage of a freighter off the coast of Scotland.

Figure 1: A superior mirage of a freighter off the coast of Scotland. The ship is actually out of sight below the horizon, but the extremely cold air refracts the light and makes the ship appear to be above the horizon and "floating" in air. Author's collection.

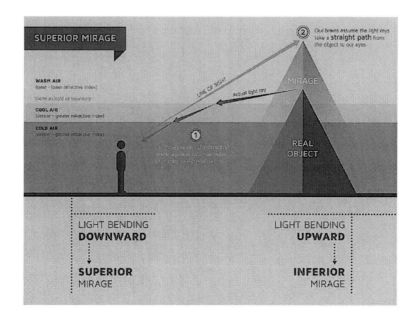

Figure 2: A graphic representation of a superior mirage. Source: Wikipedia

In the late afternoon of April 14, 1912, the Titanic had passed from the relatively warm waters of the Gulf Stream into the much colder waters of the Labrador Current. As you recall from the previous chapter, the air temperature dropped about 25° in the half-hour between 7:00 and 7:30 p.m. The combined drop in air and water temperatures would create conditions conducive to a superior mirage.

Looking again at Figure 1, it is evident that this is a superior mirage because it's daylight, and the ship is above the horizon. If the same situation were to occur at night, being able to tell that it was a superior mirage, or a mirage at all, would be much more difficult as there would be no visible horizon to provide a frame of reference.

The refraction of light that causes these mirages is much more evident at sunrise and sunset. Sunset, in particular, can cause a green flash as the atmosphere separates the different wavelengths of sunlight into its natural spectrum. Generally, the green flash won't last longer than a couple of seconds, although airline pilots flying towards the setting sun may experience a slightly longer green flash, or flashes, as their airspeed decreases the apparent speed at which the sun sets.[1] Indeed, the refraction of light causes the vast array of colors visible at sunrise and sunset.

ENDNOTES

[1] https://en.wikipedia.org/wiki/Green_flash Last accessed September 23, 2022

CHAPTER 4

The Wireless Operators

Virtually every lie starts with a kernel of truth. Let me begin this chapter by dispelling one of the most persistent lies about the Titanic sinking. The lie is that David Sarnoff (who would later go on to found RCA) operated his Marconi Wireless station at Wannamaker's Department Store in New York City for 72 hours straight. He claimed to have spent the time relaying messages from the Carpathia as it returned to New York with the Titanic survivors on board. The only bits of truth in that story are 1-David Sarnoff was an employee of the Marconi Company in 1912, and 2-his station was at the Wannamaker Department Store. April 14, 1912, was a Sunday. Wannamaker's was closed; Sarnoff's Marconi station was off the air. The station at Wannamaker's was a low-power transmitter, and there was fear that it would interfere with the operation of the more powerful Marconi stations at Cape Race, Newfoundland, Cape Cod, Massachusetts, and New Brunswick, New Jersey. Sarnoff had been told to keep his station off the air until after the Carpathia docked in New York (there is some anecdotal evidence to suggest the order may have come from Guglielmo Marconi himself). Mr. Marconi was originally going to sail to New York on the Titanic and was given a complimentary passage by the White Star Line. He switched to the Lusitania because he preferred that ship's public stenographic service and had much work to do before arriving in New York[1]. The evidence indicates that Sarnoff started this story some years after the tragedy to work himself into the history of that night. Having taken care of that bit of housekeeping, let's look at the wireless operators and their actions that night.

Morse Code uses a combination of dots and dashes to represent letters, numbers, and punctuation. A dot is one unit. A dash is three times as long as a dot. Spaces between the individual dots and dashes in one letter are one unit (a dot). Spaces between letters in words are three units (a dash), and spaces between words are seven units. There are also what were called prefixes, a (usually) three-letter combination that is analogous to "shorthand" to convey a much longer word, phrase, or "concept" and sent as one "letter." SOS (contrary to popular belief, SOS has no meaning; it's just a pattern that's easy to send …---…) and MSG (Master's Service Message—remember this one, it becomes crucial on the night of the sinking) are examples of prefixes. Figure 1 gives a complete list of prefixes used by the Marconi company. In 1912 there were two different versions, or dialects if you will, of Morse Code; American, also known as Railroad

Morse, and Continental, also known as International Morse Code, which was established in 1851.

Figure 2 graphically shows the differences between American Morse and International Morse. The differences are with the letters C, F, J, L, O, P, Q, R, X, Y, and Z, all the numbers 0-9, and the characters for the period, question mark, and comma. Of course, Morse Code has additional characters for the other punctuation marks, but this shows the basics.

On the night of April 14-15, 1912, there were many heroes. The wireless operators were undoubtedly among them. Over the roughly two-and-a-half hours it took the Titanic to sink, many wireless messages were sent and received. Yet, to the best of everyone's knowledge, only one original transcription of a Marconigram from that night is left in existence that is not in a private collection and where it is readily viewable by the public. The Weather Bureau Office received it at Cape Hatteras, NC. At some point, it was rolled up and stuffed into the wall to help insulate the building. Then, in the early 2000s, the building was undergoing a remodeling by the National Park Service into a visitor's center, and the Marconigram was discovered. It was sent to the National Park Service Conservation Center at Harpers Ferry, WV,

BI	Waiting
CQ	Calling all ships
CQD	Distress—calling all ships
DDD	Silence, shut up
DE	From, this is
G	Go ahead, start sending
K	Please reply
Nil	No message exchanged
NR	No Response
OM	Old man—a friendly form of address used between operators
P	Passenger message (commercial)
RD	Received
S	Service message (not commercial traffic)
SG/MSG	Prefix to Master's Service Message (captain must acknowledge receiving the message)
SOS	Distress (introduced in 1906)
STD BI	Please wait, stand by
TIS	End of message
TR	Time rush (exchange of signals and current time)
V	Letter V commonly sent when testing apparatus
Xs	Atmospherics which interfere with the signal

Figure 1 List of Marconi Radio Codes[2]

LETTER	AMERICAN MORSE	INTERNATIONAL MORSE	LETTER	AMERICAN MORSE	INTERNATIONAL MORSE	NUMBERS	AMERICAN MORSE	INTERNATIONAL MORSE
A	.-	.-	N	-.	-.	1	.--.	.----
B	-...	-...	O	. .	---	2	..-..	..---
C	.. .	-.-.	P	.....	.--.	3	...-.	...--
D	-..	-..	Q	..-.	--.-	4	....-	....-
E	.	.	R	. ..	.-.	5	---	.....
F	.-.	..-.	S	...	...	6	......	-....
G	--.	--.	T	-	-	7	--..	--...
H	....	....	U	..-	..-	8	-....	---..
I	..	..	V	...-	...-	9	-..-	----.
J	-.-.	.---	W	.--	.--	0	———	-----
K	-.-	-.-	X	.-..	-..-	SYMBOLS		
L	-	.-..	Y	.. ..	-.--	.	..--..	.-.-.-
M	--	--	Z	... .	--..	?	-..-.	..--..
						,	.-.-	--..--

Figure 2 A comparison of American (Railroad) Morse Code and International Morse Code.

where it was conserved, it is now on display at the Graveyard of the Atlantic Museum in Hatteras, NC, and is shown in Figure 3.

Figure 3 This Marconigram is from April 14, 1912. As far as is known, this is the only original copy of any of the distress calls from the night the *Titanic* left in existence (that is not in a private collection). The transcription reads, "Received Hatteras Station at 11:25 p.m. Titanic calling CQD giving position 41.44 N ?0.24W about 380 miles SSE of Cape Race. At 11:35 p.m., Titanic gives corrected position as 41.46 N 50.14 W. A matter of 5-or six-miles difference. He says, 'Have struck iceberg.'" Author's photo.

There is only one problem with this Marconigram. The latitude comes from a position calculated by Captain Edward J. Smith of the Titanic. The longitude comes from a position calculated by Joseph Boxhall, the Titanic's fourth officer. So, while I do not doubt the authenticity of the Marconigram, and I do not doubt that it was received at the Cape Hatteras Weather Bureau office on that fateful night, how did this Marconigram get a combination of two different navigational readings? Of all the reported transmissions, this is the only one that has this combination of latitude and longitude. Also, this Marconigram is not included in the book "*Titanic Calling Wireless Communications During the Great Disaster*," which contains records of all the Marconi transmissions that the Marconi Company compiled for the British Board of Trade Inquiry.

Working for the Marconi Company was one of the premier jobs for young men in the early 20th Century. The primary purpose of a Marconi station on a ship was as a passenger service. Although reports of weather conditions, icebergs, and other navigational information were transmitted as a courtesy for the shipping companies (and were supposed to be handled as priority messages), the Marconi company made their money by sending messages from the passengers on board the ships.

Figure 3 Jack Phillips, the Senior Marconi Operator, of the RMS Titanic, in his Marconi uniform. Photo from National Archives.

Jack Phillips, 25, was the senior Marconi operator on the Titanic. His previous assignments included the Teutonic, Lusitania, and Mauretania. He also worked for the Marconi company at several shore stations. Unfortunately, he would die in the sinking.

Harold Bride, 22, was the junior Marconi operator on the Titanic. Previously he served on board the Lusitania. He would survive, getting into Collapsible B less than 10 minutes before the Titanic sank. In 1922, Bride left the Marconi Company and went to work as a pharmaceutical salesman in Glasgow, Scotland. Suffering from what we would now call PTSD, he had grown tired of being associated with the Titanic and wanted to put some distance between himself and the memories of that night. However, he never lost interest in radio, eventually becoming an Amateur Radio operator later in life.

Figure 4 Harold Bride, the Junior Marconi Operator on the RMS Titanic, in his Junior Marconi Operator uniform. He would survive the sinking on Collapsible B. Photo from author's collection.

They boarded the ship at Belfast, Ireland. On the trip to Southampton, they worked stations in Tenerife, Canary Islands, and Port Said, Egypt[3], distances of approximately 1700 and 3500 miles, respectively. The Marconi company had estimated the transmitter's range on the Titanic to be 500 miles during the day and 1000 miles at night. In 1912, the way radio waves propagate, or travel through the atmosphere, was not completely understood; the assumption was that more power would equate to a longer range. It was not understood that the frequency also would affect the distance as well as the amount of ionization in the atmosphere. At the same time, there was no testing of the emergency transmitter.

Time for a brief physics lesson and a touch of high school algebra. $V=\lambda f$ where V=velocity (in this case, the speed of light or 299,792,458 meters per second), $\lambda=$ the wavelength, and f=the frequency. Therefore, $f=V/\lambda$ or 299,792,458/600=499.654Hz (cycles per second).

The 600-meter band (one wavelength was 600 meters or approximately 1968.5 feet long, or just under 4/10 of a mile) has a frequency of 499.654kHz, or just below the low end of the AM band (530kHz) on your radio. The 300-meter band frequency (approximate wavelength of 984.25 feet or just under 2/10 of a mile) is 999.30kHz, or almost in the middle of the AM band.

On April 13, 1912, at approximately 11:00 p.m. (GMT), the Marconi Wireless System on the Titanic broke down. The policy of the Marconi Company was to use the backup system and let a Marconi engineer repair the system when the ship arrived at its destination. The backup system on the Titanic was low power and was estimated to have a range of perhaps 50 miles. Again, neither Phillips nor Bride had tested the unit. There was a large

At about 11:00 p.m. on the night of April 14, the wireless operator from the Californian, Cyril Evans, sent out a call that they were stopped and surrounded by ice. Jack Phillips rebuked him, replying, "Shut up, shut up. I'm working Cape Race."[5] Captain Lord of the Californian did not know until after the sinking that the Titanic had rebuked the wireless call. In all fairness to Phillips though, Californian's warning did not have the MSG prefix. Nonetheless, it was a navigational

---

Figure 6 The transcription of the wireless message from the *S.S. Mesaba* to the *R.M.S. Titanic* sent at 7:50 p.m. ATS on April 14, 1912.

Message # 1
From: Mesaba
To: Titanic
Prefix: Ice Warning

Ice report in Lat 42N to 41.25N Long 49W to Long 50.30W Saw much heavy pack ice and great number large icebergs also field ice Weather good clear[4]

---

message, and according to Marconi rules, Phillips should have taken it to the bridge. Cyril Evans could and probably should have prefixed the message with the MSG prefix to avoid ambiguity.

The Titanic was one of few ships that maintained a 24-hour wireless watch. The vast majority of ships were like the Californian; one operator for a 16–18-hour shift and no specific schedule to monitor the radio. That would change after the Board of Trade hearings into the sinking.

According to Lightoller, in his book "*Titanic and Other Ships*," he recalled having the following conversation with Phillips. "Later, while standing on the upturned boat (Collapsible B), Phillips explained when I said that I did not recall any Mesaba report: 'I just put the message under a paperweight at my elbow just until I squared up what I was doing before sending it to the Bridge.' That delay proved fatal and was the main contributing cause to the loss of that magnificent ship and hundreds of lives. Had I, as Officer of the Watch, or the captain, become aware of the peril lying so close ahead and not instantly slowed down or stopped, we should have been guilty of culpable and criminal negligence."[6] The unasked question is, if this conversation occurred, why did Second Officer Charles Herbert Lightoller not bring it up at the United States Senate or British Board of Trade inquiries?

"Conflicting and contradictory information led to popular belief that Phillips possibly managed to make it to the overturned lifeboat B, which was in charge of Second Officer Lightoller, along with

Harold Bride, but Phillips did not last the night. In his New York Times interview, Bride said that a man from boat B was dead and that as he boarded the Carpathia, he saw that the dead man was Phillips. However, Bride, when testifying in the Senate Inquiry, changed his story, saying that he had only been told that Phillips died on Collapsible B and was later buried at sea from Carpathia and had not witnessed this for himself.

"In his book (*Titanic: A Survivor's Story*, originally published as *The Truth About the Titanic*), Colonel Archibald Gracie said a body was transferred from the collapsible onto boat #12 but noted that the body was not that of Phillips. He reported that when speaking with Lightoller, the Second Officer agreed that the body wasn't Phillips. In Lightoller's Senate Inquiry testimony, he says that Bride told him that Phillips had been aboard and died on the boat, but it is clear that Lightoller never saw this for himself. In Lightoller's 1935 autobiography, *Titanic and Other Ships*, he writes that Phillips was aboard Collapsible B and told everyone the position of the various ships they had contacted by wireless and when they could expect a rescue before succumbing to the cold and dying. He also claims that Phillips' body was taken aboard Boat No.12 at his insistence.

"It is clear from Gracie and other 1912 evidence that the man on the upturned collapsible who called out the names of approaching ships was Harold Bride, not Jack Phillips, as Lightoller thought in 1935. Lightoller's 1912 testimony contradicts his 1935 statements that he saw Phillips aboard B and that the body taken off the boat was Phillips. Saloon Steward Thomas Whiteley may have been Bride's source for the story; in a press interview, Whiteley claimed that Phillips had been aboard the collapsible, died, and was taken aboard Carpathia, as no other witness in 1912 claimed Phillips' body was recovered, and any source aboard Carpathia never mentioned his name as being one of the four bodies buried at sea, it's possible that Whiteley was mistaken in his identification, or that if Phillips had been aboard Collapsible B, his body wasn't recovered." [7]

Three things that need to be kept in mind about the preceding paragraphs are that they were written 23 years after the sinking of the Titanic (Lightoller's book was initially published in 1935). Lightoller had a history of "carefully" answering questions at both inquiries, doing his best to protect his employer's reputation, the White Star Line. Also, Lightoller did not mention this information at either the U.S. Senate or British Board of Trade inquiries, which leads to the question of why; if he had said it, then at least some of the responsibility would have shifted from Captain Smith and the White Star Line to the Marconi Company. The second is that there is a record here of the Mesaba's message not having the MSG prefix, which would have given it priority in getting taken to the bridge (although, as a navigational message, it should have been taken to the bridge upon receipt, per Marconi Company rules). Also, Lightoller's book was pulled shortly after publication because of the threat of a lawsuit by the Marconi Company. Lastly, earlier in the book, when Lightoller talks about the Titanic departing Southampton, "The Oceanic and St. Paul (sic) were lying moored to the wharf alongside each other. They happened to be in a position where the Titanic had to make a slight turn, which necessitated coming astern on her port engine. The terrific suction set up in that shallow water simply dragged both these great liners away from the wharf. The St. Paul broke adrift altogether, and the Oceanic was dragged off until a sixty-foot gangway dropped from the wharf into the water. It looked as if nothing could save the St. Paul crashing into the Titanic's stern—in fact, only Captain Smith's experience and resource saved her... Just as a collision seemed inevitable, Captain Smith gave the Titanic a touch ahead on her port engine, which simply washed the St. Paul away, and kept her clear until a couple of tugs, to our unbounded relief, got hold, and took her back alongside the wharf."[8] The ship that almost hit the Titanic was not the St. Paul but rather the New York. Additional contemporary accounts state nothing about Captain Smith applying forward power to the port engine to "wash the ship away." As the Titanic was within the confines of Southampton harbor when this incident occurred, the ship was not under the command of Captain Smith but rather a harbor pilot.

Additionally, as Second Officer, he (Lightoller) would have been on station in the Crow's Nest with the lookouts, per White Star Line regulations. How would he know what Captain Smith was doing on the bridge? Therefore, Lightoller's statements should again be taken with a grain of salt as it appears that he wants to give his version of the events, even when the facts contradict him, including the fact that he was nowhere near the bridge while the ship was leaving and that Phillips died during the sinking and never made it to Collapsible B. There is no record of Jack Phillips' body having ever been recovered.

ENDNOTES

[1] Smithsonian Magazine, April 1, 2012

[2] Titanic Calling Wireless Communications During the Great Disaster, First Edition, 2012, Bodleian Library, p 158

[3] On A Sea of Glass, Third Edition, 2015, Amberly Publishing, p 53

[4] Titanic Calling Wireless Communications During the Great Disaster, First Edition, 2012, Bodleian Library, p 124

[5] Stanley Lord: Captain of SS Californian (encyclopedia-titanica.org) Last accessed September 23, 2022.

[6] Titanic and Other Ships, Charles Herbert Lightoller. Chapter 31. Project Gutenberg of Australia ebook. The most recent update was in April 2005. It was last accessed on February 14, 2023. This upload does not have individually marked pages, making referencing a page number impossible.

[7] Jack Phillips (wireless officer) - Wikipedia Jack Phillips Death, Wikipedia.org. It was last accessed on December 17, 2022.

[8] Titanic and Other Ships, Charles Herbert Lightoller. Chapter 31. Project Gutenberg of Australia ebook. Most recent update April 2005. Last accessed February 14, 2023. This upload does not have individually marked pages, making referencing a page number impossible.

CHAPTER 5

Navigation

Celestial navigation is arguably the most challenging course taught at the United States Naval Academy. I hopefully have succeeded in making the subject understandable in just a few pages.

Before the advent of Global Positioning System (GPS) satellites, navigation on the open seas depended on celestial navigation. The tools of celestial navigation are simple; a sextant, a chronometer (a very accurate watch or clock with a known time deviation), and the necessary sight reduction tables; Burdwood and Böwditch are two of the more common sets of tables. A competent navigator could easily plot his position within one mile, and often accuracy to a quarter of a mile was possible. The tools used in celestial navigation are not dependent on modern computer networks; therefore, hacking into the navigation is impossible, and any errors will become immediately apparent; it is a self-correcting system.

Sighting, also called a fix or shooting the star, would be most accurate when taken just before sunrise and after sunset, when both bright stars and the horizon would be visible (Navigational Twilight). Ideally, it requires two people to sight a star accurately; one to sight it with the sextant and bring the star to the horizon, and the second to record the observation time. An experienced navigator can obtain the necessary sights in 15-20 minutes. Each sighting will give a Line of Position

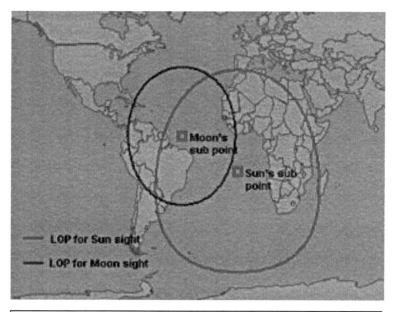

Figure 1 A set of Lines of Position. When projected onto a large map, lines of position become circles. In this example, if you were on a ship, you would be some distance west of Casablanca, Morocco. The larger circle is the Line of Position for the Sun sight, while the smaller circle is the Line of Position for the Moon sight. If they were projected onto a globe, they would be true circles.

(LOP), meaning that the ship will lie somewhere along that line. Two sightings will give an intersection which gives the ship's position. Three sightings will yield a small triangle (also called a cocked hat) containing the ship's actual position. The size of the triangle would be determined by the ship's speed and time intervals in taking the sightings. Assuming that three fixes are recorded, if a math error occurs, the LOP that is in error will be immediately evident because it will be a line that does not intersect the other two LOPs.

Figure 3 A typical modern marine sextant. It is pointed at the horizon then, by using a split mirror system, the object being sighted, or shot, is brought down to the horizon, the angle is read, and, by using tables, a Line of Position is determined, Author's collection.

It is also possible to take a sighting at night, but the accuracy would suffer because of the difficulty in determining the horizon.

Figure 6 A ship's officer demonstrating a sextant in use. When within sight of the shore, the sextant can be held horizontally to measure the angle between two landmarks and the result plotted on a nautical chart to determine your position. Author's collection.

The next logical question is how one determines their position between sights or fixes.

The answer is deducing your position, or dead reckoning, where your location from the last known fix is calculated based on your ship speed (the RPM of the propellers), direction, time since the previous fix, and allowances for current and wind.

LATITUDE 30°.

DECLINATION — *contrary Name to* — LATITUDE.

Apparent Time. A.M.	0°	1°	2°	3°	4°	5°	6°	7°	8°	9°	10°	11°	Apparent Time. P.M.
h. m.	° ′	° ′	° ′	° ′	° ′	° ′	° ′	° ′	° ′	° ′	° ′	° ′	h. m.
IX. 0	116 35	117 32	118 29	119 26	120 21	121 16	122 9	123 2	123 54	124 45	125 36	126 26	III. 0
VIII. 56	115 48	116 44	117 41	118 38	119 33	120 28	121 21	122 14	123 6	123 57	124 48	125 38	4
52	115 1	115 57	116 54	117 51	118 46	119 41	120 34	121 27	122 19	123 10	124 1	124 51	8
48	114 15	115 11	116 8	117 5	118 0	118 55	119 48	120 41	121 33	122 24	123 15	124 5	12
44	113 30	114 26	115 23	116 20	117 15	118 9	119 2	119 55	120 47	121 39	122 30	123 20	16
40	112 46	113 42	114 39	115 35	116 30	117 24	118 17	119 10	120 2	120 54	121 45	122 35	20
36	112 3	112 59	113 55	114 51	115 46	116 40	117 33	118 26	119 18	120 10	121 1	121 51	24
32	111 21	112 17	113 12	114 8	115 3	115 57	116 50	117 43	118 35	119 27	120 18	121 8	28
28	110 39	111 35	112 30	113 26	114 21	115 15	116 8	117 1	117 53	118 45	119 36	120 26	32
24	109 58	110 54	111 49	112 45	113 39	114 33	115 26	116 19	117 11	118 3	118 54	119 44	36
20	109 18	110 14	111 9	112 4	112 58	113 52	114 45	115 38	116 30	117 22	118 13	119 3	40
16	108 38	109 34	110 29	111 24	112 18	113 12	114 5	114 58	115 50	116 41	117 32	118 23	44
12	107 59	108 55	109 50	110 45	111 39	112 32	113 25	114 18	115 10	116 1	116 52	117 43	48
8	107 21	108 16	109 11	110 6	111 0	111 53	112 46	113 39	114 31	115 22	116 13	117 4	52
4	106 43	107 38	108 33	109 28	110 22	111 15	112 8	113 0	113 52	114 43	115 34	116 25	56
VIII. 0	106 6	107 1	107 56	108 50	109 44	110 37	111 30	112 22	113 14	114 5	114 56	115 47	IV. 0
VII. 56	105 29	106 24	107 19	108 13	109 7	110 0	110 52	111 44	112 36	113 27	114 18	115 9	4
52	104 53	105 48	106 42	107 36	108 30	109 23	110 15	111 7	111 59	112 50	113 41	114 32	8
48	104 17	105 12	106 6	107 0	107 53	108 46	109 38	110 30	111 22	112 13	113 4	113 55	12
44	103 42	104 37	105 31	106 24	107 17	108 10	109 2	109 54	110 46	111 37	112 28	113 19	16
40	103 7	104 2	104 56	105 49	106 42	107 34	108 26	109 18	110 10	111 1	111 52	112 43	20
36	102 33	103 27	104 21	105 14	106 7	106 59	107 51	108 43	109 35	110 26	111 17	112 8	24
32	101 59	102 53	103 46	104 39	105 32	106 24	107 16	108 8	109 0	109 51	110 42	111 33	28
28	101 25	102 19	103 12	104 5	104 58	105 50	106 42	107 34	108 26	109 17	110 8	110 59	32
24	100 52	101 45	102 38	103 31	104 24	105 16	106 8	107 0	107 52	108 43	109 34	110 25	36
20	100 19	101 12	102 5	102 58	103 50	104 42	105 34	106 26	107 18	108 9	109 0	109 51	40
16	99 46	100 39	101 32	102 25	103 17	104 9	105 1	105 53	106 45	107 36	108 27	109 18	44
12	99 14	100 6	100 59	101 52	102 44	103 36	104 28	105 20	106 12	107 3	107 54	108 45	48
8	98 42	99 34	100 27	101 19	102 11	103 3	103 55	104 47	105 39	106 30	107 21	108 12	52
4	98 10	99 2	99 55	100 47	101 39	102 31	103 23	104 15	105 7	105 58	106 49	107 40	56
VII. 0	97 38	98 30	99 23	100 15	101 7	101 59	102 51	103 43	104 35	105 26	106 17	107 8	V. 0
VI. 56	97 6	97 58	98 51	99 43	100 35	101 27	102 19	103 11	104 3	104 54	105 45	106 36	4
52	96 35	97 27	98 20	99 12	100 4	100 55	101 47	102 39	103 31	104 22	105 13	106 4	8
48	96 4	96 56	97 49	98 41	99 33	100 24	101 16	102 8	103 0	103 51	104 42	105 33	12
44	95 33	96 25	97 18	98 10	99 2	99 53	100 45	101 37	102 29	103 20	104 11	105 2	16
40	95 2	95 54	96 47	97 39	98 31	99 22	100 14	101 6	101 58	102 49	103 40	104 31	20
36	94 31	95 24	96 16	97 8	98 0	98 51	99 43	100 35	101 27	102 18	103 9	104 0	24
32	94 0	94 53	95 45	96 37	97 29	98 20	99 12	100 4	100 56	101 47	102 39	103 30	28
28	93 30	94 22	95 14	96 6	96 58	97 50	98 42	99 34	100 26	101 17	102 9	103 0	32
24	93 0	93 52	94 44	95 36	96 28	97 20	98 12	99 4	99 56	100 47	101 39		36
20	92 30	93 22	94 14	95 6	95 58	96 50	97 42	98 34	99 26			Sun *rises* $6^h 26^m$ *sets* $5^h 34^m$ bears 102° 44′	40
16	92 0	92 52	93 44	94 36	95 28	96 20	97 12			Sun *rises* $6^h 21^m$ *sets* $5^h 39^m$ bears 100° 24′	Sun *rises* $6^h 23^m$ *sets* $5^h 37^m$ bears 101° 34′		44
12	91 30	92 22	93 14	94 6	94 58			Sun *rises* $6^h 16^m$ *sets* $5^h 44^m$ bears 98° 5′	Sun *rises* $6^h 19^m$ *sets* $5^h 41^m$ bears 99° 15′				48
8	91 0	91 52	92 44	93 36		Sun *rises* $6^h 12^m$ *sets* $5^h 48^m$ bears 95° 47′	Sun *rises* $6^h 14^m$ *sets* $5^h 46^m$ bears 96° 56′						52
4	90 30	91 22	Sun *rises* $6^h 5^m$ *sets* $5^h 55^m$ bears 92° 19′	Sun *rises* $6^h 7^m$ *sets* $5^h 53^m$ bears 93° 28′	Sun *rises* $6^h 9^m$ *sets* $5^h 51^m$ bears 94° 37′								56
VI. 0	90 0	Sun *rises* $6^h 2^m$ *sets* $5^h 58^m$ bears 91° 10′											VI. 0

In North Latitude { *When Apparent Time is A.M. read the Azimuth from North to East.*
" ——— *P.M.* ——— " ——— *North to West.*

Figure 4 A page from the Burdwood Table, circa 1909. Public domain.

TABLE 2. [Page 611 Difference of Latitude and Departure for 41° (139°, 221°, 319°). Dist. Lat. Dep. Dist. Lat. Dep. Dist. Lat. Dep. Dist. Lat. Dep. Dist. Lat. Dep. 1 0.8 0.7 61 46.0 40.0 121 91.3 79.4 181 136. 6 118. 7 241 181.9 158. 1 2 1.5 1.3 62 46.8 40.7 22 92.1 80.0 82 137. 4 119. 4 42 182.6 158.8 3 2.3 2.0 63 47.5 41.3 23 92.8 80. 7 83 138. 1 120.1 43 183.4 159. 4 4 3.0 2.6 64 48.3 42.0 24 93.6 81.4 84 138. !I 120. 7 44 184.1 160. 1 5 3.8 3.3 65 49.1 42.6 25 94.3 82.0 85 139. Ii 121.4 45 184.9 160. 7 6 4.5 3.9 66 49.8 43.3 26 95. 1 82.7 86 140. 4 122.0 46 185.7 161.4 7 5.3 4.6 67 50.6 44.0 27 95.8 83.3 87 141. 1 122.7 47 186.4 162. 0 8 6.0 5.2 68 51.3 44.6 28 96.6 84.0 88 141. !> 123. 3 48 187.2 162.7 9 6.8 5.9 69 52. 1 45.3 29 97.4 84.6 89 142.)i 124.0 49 187.9 163. 4 10 7.5 6.6 70 52.8 45.9 30 98.1 85.3 90 143. 4 124.7 50 188.7 164.0 11 8.3 7.2 71 53.6 46.6 131 98.9 85.9 191 144. L 125.3 251 189.4 164.7 12 9.1 7.9 72 54.3 47.2 32 99.6 86.6 92 144. 9 126.0 52 190. 2 165. 3 13 9.8 8.5 73 55.1 47.9 33 100.4 87.3 93 145. 7 126.6 53 190.9 166.0 14 10. 6 9.2 74 55.8 48.5 34 101.1 87.9 94 146. o 127.3 54 191. 7 166. 6 15 11.3 9.8 75 56.6 49.2 35 101.9 88.6 95 147. 2 127.9 55 192.5 167.3 16 12.1 10.5 76 57.4 49.9 36 102. 6 89. 2 96 147. i) 128.6 56 193.2 168. 0 17 12.8 11.2 77 58.1 50.5 37 103.4 89.9 97 148. 7 129:2 57 194.0 168. 6 18 13.6 11.8 78 58.9 51.2 38 104. 1 90.5 98 149. 4 129.9 58 194.7 169.3 19 14.3 12.5 7!) 59.6 51.8 39 104.9 91.2 99 150. 2 130.6 59 195. 5 169. 9 20 15.1 13.1 80 60.4 52.5 40 105.7 91.8 200 150. !I 131.2 60 196.2 170.6 21 15.8 13.8 81 61. 1 53.1 141 106.4 92. 5 201" 151. 7 131".9 261 197. Ö" 171".2" 22 16.6 14.4 82 61.9 53.8 42 107.2 93.2 02 152. 5 132.5 62 197.7 171.9 23 17.4 15.1 83 62.6 54.5 43 107.9 93.8 03 153. 2 133.2 63 198.5 172.5 24 18.1 15.7 84 63. 4 55. 1 44 108.7 94.5 04 154. 0 133.8 64 199.2 173. 2 25 18.9 16.4 85 64.2 55.8 45 109. 4 95. 1 05 154. 7 134.5 65 200.0 173.9 26 19. 6 17. 1 86 64.9 56.4 46 110.2 95.8 06 155. 5 135. 1 66 200. 8 174.5 27 20.4 17.7 87 65. 7 57. 1 47 110.9 96.4 07 156. 2 135.8 67 201.5 175. 2 28 21.1 18.4 88 66. 4 57. 7 48 111.7 97. 1 08 157. 0 136. 5 68 202. 3 175.8 29 21.9 19.0 89 67.2 58.4 49 112.5 97.8 09 157. 7 137.1 69 203.0 176.5 30 22.6 19.7 90 67.9 59.0 50 113.2 98.4 10 158. 5 137.8 70 203. 8 177.1 31 23.4 20.3 91 *~68. 7" 59.7 151 114.0 99. 1 211 159. o> 138. 4 271" 204.5 177.8 32 24.2 21.0 92 69. 4 60.4 52 114.7 99.7 12 160. 0 139. 1 72 205. 3 178.4 33 24.9 21.6 93 70.2 61.0 53 116. 5 100.4 13 160.8 139.7 73 206.0 179. 1 34 25.7 22. 3 94 70.9 61.7 54 116.2 101.0 14 161. 5 140. 4 74 206.8 179.8 35 26.4 23.0 95 71.7 62.3 55 117.0 101. 7 15 162. 3 141. 1 75 207.5 IS0. 4 36 27.2 23.6 96 72.5 63.0 56 117.7 102.3 16 163. 0 141. 7 76 208. 3 181. 1 37 27.9 24.3 97 73.2 63.6 57 118.5 103. 0 17 163. 8 142.4 77 209. 1 181.7 38 28.7 24.9 98 74.0 64.3 58 119.2 103.7 18 164. 5 143. 0 78 209.8 182.4 39 29.4 25.6 99 74.7 64.9 59 120.0 104.3 19 165. 3 143.7 79 210.6 183.0 40 30.2 26.2 100 75. 5 65. 6 60

Figure 5 A portion of a Bowditch Table, circa 1900. Public domain.

The ship's slip is the difference between the speed of the engine and the actual observed speed of the ship. Slip = 100% - Efficiency, where efficiency = the observed distance covered or engine speed. For example, during a 24-hour period, a ship's propeller shaft produced 72 revolutions per minute (RPM), the propeller's pitch was 4 meters, and the observed speed over the ground was 10 knots. A nautical mile equals 1852 meters. To calculate the value of the propeller slip:

Slip % = (Engine distance - ship's distance) / (engine distance x 100)

Engine distance = Pitch x RPM x 60 x 24 / 1852

= 4 x 72 x 60 x 24 / 1852

= 414,720 / 1852

=223.930

Ship's distance = 24 x 10

= 240

Therefore slip = (223.930-240)/(223.930 x 100) or 7.17%.[1,2]

A slip table is one of the first things created once a ship enters service. The slip table allows for a direct conversion of RPM into speed. As the

Titanic was on her maiden voyage, the slip table was in the process of being created as it did not yet exist, so the speeds quoted were, at best, estimates. However, the existing data suggest that they were very accurate estimates. Also, depending on the strength of the ocean currents the ship would encounter, there could be two slip tables, one for eastbound trips and one for westbound trips.

Sailors must resort to dead reckoning their position when they don't have current sights. Dead reckoning requires that the navigator interpolate or deduce the ship's position. Deducing your position requires the position as determined by the last fix, the heading, speed (hence the need for a slip table), a chronometer so you know how much time has passed, and knowledge of the winds and currents that the ship is passing through.

Dead reckoning will generally introduce cumulative errors in determining the ship's position as the effects of current, and to a lesser degree, the wind, are estimated based on the officer's experience. Several days of bad weather could lead to a significant cumulative error.

In the Titanic's case, the weather was remarkably clear for most of the voyage, allowing for accurate fixes at dawn and dusk (Navigational Twilight). During the navigational fixes, the dead reckoning positions were not too far off from the actual positions.

This begs the question of how there was a 13-mile difference between the Titanic's actual position (based on the wreck's position) and the position the wireless operators were sending out. The positions given in the Marconigram shown in Chapter 4, Figure 2 don't result in a 13-mile error, but they aren't accurate either. As I speculated in Chapter 3, (The Weather), with the sea being so smooth, it would be difficult to get a precise fix on the horizon after dark. That difficulty, plus the ship's sinking, would reduce the distance to the visible horizon, affecting the calculations determining the Titanic's latitude and longitude. Another factor of greater importance is the correct time, which will be addressed in the next chapter.

Figure 7 shows a compass card from about 1912. It has two points of zero degrees (at North and South) and two points of 90 degrees (at East and West). The command, "North, 45 degrees East" would result in a heading to the Northeast. Likewise, the command "South, 67.5 degrees West" would put the ship on a West-Southwest heading. That is how all commands to steer ships were given until the advent of the gyro compass.

Figure 7 A compass card, circa 1912. You will observe that both North and South have readings of 0 degrees, while East and West both have readings of 90 degrees. This card predates the modern gyroscopic compass, which goes through the whole 360 degrees. Author's collection.

Figure 8 shows a modern compass card with a full 360-degree set of headings. It also has the traditional quadrants on the inner ring of headings.

Figure 8 A modern Compass Rose. The outer ring gives the headings from Zero to 360 degrees, while the inner ring shows the headings broken down into the traditional quadrants. Author's collection.

Gyro compasses began to come into use starting in 1909. A gyro compass uses a gyroscope to determine true North via gyroscopic precession. Gyro compasses, because they do not rely on magnets, are perfect for steel hull ships as the navigator doesn't have to determine the magnetic deviation caused by the ship's hull.[3]

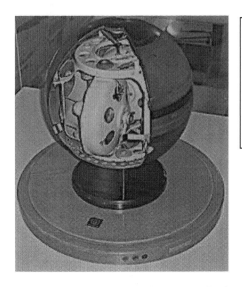

Figure 9 A cutaway view of an Anschütz Gyrocompass. Wikipedia, last accessed September 23, 2022.

With the advent of the gyro compass, the division of the Compass Rose into quadrants was unnecessary, and compasses were read from zero degrees (North) through 90 degrees (East), 180 degrees (South), 270 degrees (West), back to zero (or 360 degrees, North). With a gyrocompass, the command "North, 45 degrees East" would be "Heading 45 degrees," and the "South, 67.5 degrees West" would be "Heading 247.5 degrees (180 degrees plus 67.5 degrees)."

As previously mentioned, because a gyro compass seeks true North, the mariner does not need deviation tables or other equipment to compensate for all the steel in their ship. Also, true North is more accurate for navigation than magnetic North. This is because you do not have to compensate for the magnetic declination as you travel along the globe. Also, true North is a constant while magnetic North varies as time goes by.[4]

Most modern gyrocompasses display their readout via a compass card or a digital display.

ENDNOTES

[1] https://oic-wreviewer.blogspot.com/2014/05/find-slip-of-propeller.html

[2] https://www.youtube.com/watch?v=qTfYbl9E5l4

[3] Wikipedia article "Gyrocompass" last accessed September 13, 2022

[4]

https://en.wikipedia.org/wiki/North_magnetic_pole

Last accessed September 23, 2022

CHAPTER 6
The Importance of Time to Celestial Navigation

For celestial navigation, 59 astronomical objects are used for the required calculations (the Sun, Moon, and 57 stars). While latitude (the distance north or south of the Equator) can be calculated without knowing the time, calculating the longitude, because it is based on how far east or west of Greenwich, England, (0° longitude) the observer is located, requires knowing "when" in time the observer is. "When," in this case, refers to the time difference between the observer's position and Greenwich, England.

An example can probably best illustrate this. Assume that it is midnight where you are and that, at the same time, it is 5:00 a.m. GMT (Greenwich Mean Time). The earth rotates from West to East 360° in one day. Since there are 24 hours in one day, each hour represents 15° (360°/24 hours = 15° per hour). There is a five-hour difference between your location and Greenwich (GMT). That means that you are 75° from Greenwich, and since you are ahead of Greenwich, you are east of Greenwich.

Celestial navigation requires that a clock must be highly accurate. It wasn't until 1730, when

John Harrison, a self-taught carpenter and watchmaker, invented a marine chronometer with an at-sea accuracy greater than 1 second per month, that it was finally felt that timepieces could be accurate enough for use at sea.[1]

Before Harrison invented the marine chronometer, the lunar distance was used. This involved measuring the angle between the moon and another celestial body near the ecliptic (the apparent path of the Sun against the heavens throughout the course of a year.) "At that moment, anyone on the surface of the earth who can see the same two bodies will, after correcting for parallax (the apparent shift against background stars based on the difference of the individual's locations), observe the same angle." The navigator then consults a prepared table of lunar distances and the times they will occur. By comparing the corrected lunar distance with the tabulated values, the navigator finds the Greenwich time for that observation. Knowing Greenwich time and local time, the navigator can work out longitude.[2]

In tropical latitudes, an error in time of just 1 second can result in a position error of 500 meters or .25 nautical miles. At 45° north or south, that same one-second error can result in an error of .5 nautical miles. This is because, as you get further from the equator, the lines of longitude begin to converge or come together. If you were to stand at either pole, you would be in all 24 time zones simultaneously.

There is a quote attributed to Albert Einstein that is appropriate when talking about time and the Titanic: "Time is what the clock says."[3] And, as things unfolded during the sinking of the Titanic, it became apparent that time is what the clock says, especially when the clocks had different times. Unless otherwise listed in the endnotes, all quotations and information in this chapter comes from A Hypothesis of Times Gone Wrong (encyclopedia-titanica.org) by Brad Payne, last accessed on October 13, 2022.

There are 24 hours in a day and 360° in a circle. Dividing 360° by 24 gives 15° of longitude per hour. There are 75° of longitude between Greenwich, England, the Prime Meridian or 0° longitude, and New York (75° West longitude), meaning a five-hour difference (75/15=5). Therefore, New York was five hours behind Greenwich or GMT-5 hours. So far, this coincides with how we understand time on land.

At sea, however, things are a bit different. While on land, a new day starts at midnight; at sea, a new day begins at noon. The time changes on ships during the night, as this will have less of an impact on the passengers and most of the crew. On the White Star Line ships, the time change was to occur between 10 p.m. and 6 a.m.

The Titanic had two master clocks, each controlling 24 slave clocks. The time change would generally occur around midnight as there was a shift change at that time, so half the required time change would be applied before and half after the change; this was reflected in the testimony at the British Board of Trade hearing.

While in their home port or destination, the time on the ship would coincide with the time on land. However, when a ship is at sea, things become "different." To allow for this, at sea time is recorded as Apparent Ship Time or ATS. The acronym AST could not be used because it was already in use for Atlantic Standard Time (the time zone at the 60° W longitude meridian or GMT-4 hours). Ships A and B are both traveling west and A passes B. If you could freeze the moment when both ships are side by side, you would find that, although the chronometers set to GMT both displayed the same time, the other clocks, controlled by the ship's master clock(s), could be off by maybe as much as five to ten minutes between the two ships. This reflects what time midnight occurred on board each vessel.

Let me illustrate this point as follows. Both ships are at 45° North latitude. That means that each minute of longitude is half of a nautical mile. Ship A is at 37° 15' West when her clocks are reset to midnight. Ship B is at 38° 15' West when her clocks are reset to midnight. The ships are separated by one degree, which equals 60 minutes of longitude. The ships are 30 miles apart (60 minutes x .5 nautical miles = 30), and the time difference between the ships would be about two minutes (60 minutes in 15 degrees of longitude. 120/60=2 minutes). Therefore, at the instant when ship A passes ship B, the ATS differs between the two ships by two minutes. Einstein was right; time is what the clock says, and that quote becomes all the more prevalent when it comes to what time events occurred during the Titanic's sinking.

It is generally accepted that the Titanic collided with the iceberg at 11:40 p.m. ATS on April 14 and sank at 2:20 a.m. ATS on the morning of April 15, giving an elapsed time of 2 hours and 40 minutes for the great ship to sink.

But what times did the survivors give for the collision?

Trimmer Dillon	11:35 p.m.
Fireman Hurst	11:20 p.m.
Bath Steward Widgery	11:35 p.m.
Baker Joughin	11:40 p.m.
2nd Class Passenger Cook	11:25 p.m.
1st Class Passenger Hoyt	11:35 p.m.
3rd Class Passenger Jansson	11:30 p.m.
1st Class Passenger Lines	11:45 p.m.

Likewise, there were variances as to the time of the sinking.

Storekeeper Prentice	1:30 a.m.
Barber Weikman	1:50 a.m.
Seaman Scarrott	2:20 a.m.
2nd Officer Lightoller	2:30 a.m.
1st Class Passenger Leader	3:00 a.m.
2nd Class Passenger Hamalainen	2:10 a.m.
1st Class Passenger Dick	2:20 a.m.

The collision reports occur within 25 minutes, between 11:20 p.m. and 11:45 p.m. The reports of the sinking time cover a much more extended time, 1½ hours, from 1:30 a.m. to 3:00 a.m. In each case, there are four crew members of the ship listed, people who could be considered trained observers. They had, to varying degrees, knowledge of how the ship operated.

There is a time constant, as noon began a new day on the *Titanic*, that was used to establish the time relative to New York and England. When it was noon ATS on April 14, 1912, on the *Titanic,* the ship was 2 hours 58 minutes behind England, and therefore 2 hours 2 minutes ahead of New York (2 hours 58 minutes plus 2 hours 2 minutes equals 5 hours, the time difference between England and New York). Therefore, when it was noon ATS on the Titanic, it was 2:58 p.m. in England and 9:58 a.m. in New York. That time difference would remain until noon on April 15, when the next sighting would have occurred and the process repeated.

Quartermaster Hichens stated at one of the hearings, "The clock was to go back that night (April 14) 47 minutes, 23 minutes in one watch and 24 in the other." We also know from Hichens that he was relieved at 12:23 a.m., which means he received the 23-minute change. Even though the clocks were changed around midnight, that should not affect the ATS on board the Titanic, as that time change would not be calculated until noon on April 15, 1912.

Yet we have the following time differences between the Titanic and New York or London, as reported in <u>A Hypothesis of Times Gone Wrong</u>

(encyclopedia-titanica.org). Fourth Officer Boxhall would state that 11:46 p.m. (the time that he calculated the Titanic's distress coordinates would equate to 10:13 p.m. in New York, a difference of 1 hour and 33 minutes ahead of New York, not 2 hours and 2 minutes as determined by the noon ATS. Boxhall's time creates a time difference of 29 minutes between ATS at the time of the collision and ATS calculated at noon.

Lightoller would state that 2:20 a.m. (the time the Titanic sank) would equate to 5:47 a.m. GMT, a difference of 3 hours and 27 minutes. This conveniently agrees with Boxhall's time difference of 1 hour and 33 minutes (1 hour 33 minutes + 3 hours 27 minutes = 5 hours, the difference between GMT and New York time). Remember these figures; they are referenced later on.

To make things more interesting, the British Board of Trade inquiry would fix the ship's time as 1 hour and 50 minutes ahead of New York. In comparison, the Limitation of Liability report would fix the Titanic's time as 1 hour and 39 minutes ahead of New York. However, neither report states how those times are derived.

Because the Earth rotates 15° per hour, a four-second error in the time figure will result in a position error of up to one nautical mile (at the equator) or an error of ½ nautical mile at 45° north or south. Longitude is the only measurement affected by time because time is necessary for calculating how far East or West you are of a given point (generally the Prime Meridian, or 0° longitude).

Add into this the errors of the distress coordinates. Captain Smith's coordinates are 41° 44' N, 50° 24' W, or 20 miles west of the boiler field, which is the accepted position of the wreck as the boilers were so heavy, they would most likely have sunk straight down when the ship broke apart. Smith's numbers would have required almost an extra hour of steaming, based on the Titanic's assumed speed of 21½ knots. As Captain Smith was highly experienced, making such an error is inexplicable.

Boxhall's distress coordinates are just as puzzling. He arrived at 41° 46' N 50° 14' W, or some 13 miles west of the boiler field. This would have required the ship to steam an extra 35 minutes. (To refresh your memory, the Marconigram transcript referenced in Chapter 4 gives the coordinates as 41° 44' N 50° 14' W Capt. E.J. Smith's latitude and Fourth Officer Boxhall's longitude.) Boxhall's clock error is approximately four minutes and fifteen seconds.

All told, there are at least six different times here. Is there a way to reconcile these time discrepancies that could explain these differences and the navigational errors of both Captain Smith and Fourth Officer Boxhall?

Again, we must turn to [A Hypothesis of Times Gone Wrong (encyclopedia-titanica.org)](encyclopedia-titanica.org) to explain what could have occurred. The Magneta Company manufactured the Titanic's synchronized clock system. Let's start by explaining how the clock system worked or was thought to work, on the Titanic.

Synchronized clocks have a master clock and a number of slave clocks. They were electrically connected, and as each minute passed, the master clock would send a signal to the slave clocks to advance in time by one minute. Depending on the particular model, the master clock may or may not have had a second hand, but none of the slave clocks had second hands. There were also two different electrical impulses; one for odd minutes and one for even minutes.

GMT	2:35	2:38	2:44	2:58	3:04
NYC	9:35	9:38	9:44	9:58	10:04
MC-1	11:37	11:37	11:37	11:37	11:37
MC-2	11:37	11:40	11:46	12:00	12:06

GMT	3:05	3:12	3:13	3:25	5:18
NYC	10:05	10:12	10:13	10:25	12:18
MC-1	11:38	11:45	11:46	11:38	1:51
MC-2	12:07	12:14	12:15	12:27	2:20

Figure 2 A spreadsheet illustrating what Brad Payne believes may have happened regarding the time adjustments to Master Clocks 1 and 2 on the *Titanic* on the night of April 14, 1912. The time differences introduced here can offer a reasonable explanation for the navigational errors made by Captain Smith and Fourth Officer Boxhall, as well as accounting for most of the variously reported sinking times.

While this hypothesis doesn't answer all the questions, and they likely will never be answered, it gives a theoretical starting point. It must also be stated that there are those within the Titanic community that there must be other, more straightforward explanations to explain the time differences.

ENDNOTES

[1] John Harrison - Wikipedia Last accessed November 15, 2022.

[2] Lunar distance (navigation) - Wikipedia Last accessed November 15, 2022.

[3] www.azquotes.com

CHAPTER 7
Lifeboats

The Titanic did not have enough lifeboats for all the passengers and crew. At maximum capacity, the Titanic would have 3,547 passengers and crew. On the maiden voyage, the ship carried approximately 2,208 passengers and crew. The 20 lifeboats could hold 1,178 people, or a little more than half the passengers and crew on board at the time of the sinking. The British Board of Trade regulations in 1912 would require a ship the size of the Titanic to have 16 lifeboats, so the Titanic had four more boats than the law required. The critical flaw is that the law was based on the ship's gross tonnage instead of the number of passengers she could carry.

What must be remembered is that, before the Titanic's sinking in 1912, lifeboats were thought of more as "ferries" than rescue vehicles for long-term occupation; a distressed ship would call for assistance. The rescue ship(s) would arrive, and the passengers would ferry from the sinking ship to the rescue vessel(s) in the lifeboats. Part of this mindset was due to the large number of ships that traveled the established shipping lanes on the world's oceans. Also, it was thought that any ship taking on water would stay afloat long enough for rescue ship(s) to arrive, as happened with the sinking of the SS Republic in 1909. The ship was mortally wounded in a collision yet remained afloat long enough for all her passengers to be transferred to the SS Florida, using lifeboats from both ships (see pages 109-111).

It has been postulated that had there been enough lifeboats; probably everyone could have been saved. There is a flaw in that logic. The first was Second Officer Lightoller's strict interpretation of the captain's order of "Women and children first." Lightoller interpreted it to mean "Women and children ONLY." Would Lightoller's interpretation of the order have been different had there been enough lifeboats? We have no way of knowing that answer. The only men he let onto the lifeboats on the ship's port side were crewmen to row the boats and an officer in charge. Lightoller did allow one civilian man to board a lifeboat; Major Arthur Peuchen, the Canadian yachtsman, would be asked to descend the falls (the lines lowering the lifeboat from the davit) to man lifeboat #6 later that night when it was discovered the lifeboat was shorthanded. On the starboard side of the ship, First Officer Murdoch allowed men to board the lifeboats once no women or children were in the area. Third, most boats were being lowered without being near full. There was a fear that, had the boats been filled to capacity, the davits would not be able to hold the weight, although that fear was unfounded. Not until the end were boats lowered, filled to, or sometimes exceeding their capacity. It was a case of the crew not knowing the design parameters of the Welin davits.

But the most pressing issue was time. When Captain Smith met with Thomas Andrews, the ship's designer, at 12:25 a.m., Andrews said that the Titanic would sink in an hour, maybe an hour and a half (in fact, the Titanic sank at 2:20 a.m., 1 hour and 55 minutes later). If the Titanic had the full complement of lifeboats (64 boats), they would have had to launch one lifeboat every two minutes, a physical impossibility. Assuming that all the lifeboats were being filled to capacity, it would still require about 40 lifeboats, or launching one boat every four minutes, again a physical impossibility. Of the 20 lifeboats on board, the final two (Collapsibles A and B) floated off the ship just moments before the Titanic began her final plunge.

There had been no lifeboat drills except one done the morning of the ship's sailing to satisfy the inspectors. The design of the davits would have allowed the lifeboats to be lowered fully loaded with passengers without having to worry that a boat could split in two while it was being lowered. In other words, there was a failure to communicate the upgrades that had been done to these systems.

The Titanic was equipped with a new design by Welin Davits. This design allowed for multiple lifeboats to be stored at each davit location. The early drawings for the Titanic showed multiple lifeboats at each davit. This was necessary to get the approval for the davits to be used. It was not indicative that the Titanic would have more than the 16 required lifeboats.

Interestingly, Thomas Andrews (the Titanic's designer) sided with the Welin Company on using the new davits. White Star Line opposed the idea of multiple boats at each davit. I'm not saying that safety was of no concern, but they believed that the ship was unsinkable, and even if the ship were mortally wounded, she would stay afloat long enough for another ship to come and rescue everyone.

There's also the question of how the Titanic's musicians affected the evacuation proceedings. The musicians played light, cheerful music for most of the evening. The argument has been made that conflicting messages were being sent psychologically. Ship officers were ordering, sometimes begging, passengers into the lifeboats, while the musicians were playing music designed to calm passenger's potential fears. Psychologists would say that mixed messages were being sent, and they would be right. Your ship is sinking, and you're being asked to get into a tiny lifeboat. Yet the music is urging you to stay on board the larger ship because of the ship's apparent safety.

It wasn't until the end that the hymn "Nearer My God to Thee," was played, and by then, it was too late. There have been some questions as to whether "Autum" or "Nearer My God to Thee" was the last hymn played before the Titanic sank. Bandleader Wallace Hartley had been asked earlier in his career if he were on a sinking ship, what hymn would he like to hear, and his reply was "Nearer My God to Thee." On that basis, I would assume that "Nearer My God to Thee" was the last hymn played before the Titanic sank. Some have gone so far as to say that Wallace Hartley and the rest of the Titanic's band bear some responsibility for the 1496 deaths that occurred during the sinking. I wouldn't go that far, but I would say that their actions did reduce the panic while the lifeboats were being loaded, and in that regard, they probably saved some lives.

Like the wireless operators, the musicians were in a no man's land. They were not crew members but were employees of an independent company, C. W. & F. N. Black of Liverpool. This company supplied virtually all British ships with musicians. They all boarded at Southampton and traveled as second-class passengers. All would perish in the sinking.

CHAPTER 8

J. Bruce Ismay

No book about the Titanic and the events during the sinking would be complete without a chapter about the Chairman and Managing Director of the White Star Line, Joseph Bruce Ismay. The sinking of the Titanic is one of the most infamous tragedies in history, and much of the blame has been placed on J. Bruce Ismay. Some believe he was a coward who abandoned the ship, while others argue he was a hero who helped save lives. But who was J. Bruce Ismay? In this chapter, we'll delve into Ismay's life and legacy, separating fact from fiction and debunking the myths surrounding his involvement in the Titanic disaster. From his upbringing as the son of a shipping magnate to his controversial decisions as managing director of the White Star Line, we'll explore the complex character of J. Bruce Ismay and attempt to answer the question: hero or villain? So, sit back, grab a lifejacket, and let's set sail on this fascinating part of the Titanic's story.

Ismay was an interesting person. He was somewhat shy and, like most shy people, tried to keep a "barrier" up to protect himself; as a way to keep from getting hurt. Yet, if you took the time to get to know him, to break down that "barrier," you would discover a friendly individual. He was a person who chose his friends carefully. There is also some evidence to suggest that he was not entirely comfortable with public speaking.

Some 20 years before the Titanic disaster, Ismay was the White Star Line's agent in New York City. During that time, he had met William Randolph Hearst, who developed an almost instant dislike for Ismay. The feeling was mutual on Ismay's part. If Hearst hated you, he would wait as long as it took for something to happen to you so he could set his yellow journalists free on your character assassination. Therefore, the Titanic disaster provided the perfect vehicle for Hearst to perform character assassination against Ismay.

Hearst did not outright lie about Ismay, but his reporters would carefully spin and exaggerate the truth to cast Bruce Ismay in the worst light possible, trying (and succeeding) in casting him as a personification of Snidely Whiplash.

When the ship struck an iceberg on April 14, 1912, Ismay was among the first to be informed of the severity of the situation. According to some accounts, Ismay immediately took charge of the evacuation efforts and helped coordinate the rescue of passengers. However, other reports paint a different picture, suggesting that Ismay was more concerned with his own safety than that of others. Allow me to point out just one example from the Titanic. It occurs around lifeboat number five where, thanks to the yellow journalism of the time, we had variations of this exchange immortalized in the various Titanic movies:

Ismay—you must get these lifeboats loaded and the passengers away from the ship!
Third Officer Pitman—I only take orders from the captain.
Ismay—Do you know who I am?
Pitman—You are a bloody ship's passenger, and I am a ship's officer.
Then Ismay slinks away.

That makes for a good scene in a movie. Still, what really happened? Fortunately for us, Third Officer Pitman survived the sinking and relayed what happened in his testimony at the British Board of Trade Inquiry:

"At this point, while working on Lifeboat number 5, a man in his 'dressing gown and pajamas on' approaches and urges (me) to start loading the lifeboats immediately, but I refused to take his orders. At some later point, I realized that this man was Bruce Ismay, Chairman of the White Star Line."

Third Officer Pitman recalls, "Mr. Ismay assisted in helping women and children into lifeboat number five and that he stepped back as the lifeboat was lowered." Similar statements from other officers and passengers corroborate similar actions by Mr. Ismay at other lifeboats over the next two hours.

The sworn testimony of Third Officer Pitman certainly doesn't paint the same picture of the man that Hearst's "yellow journalists" do. However, Hearst's version makes for better reading, especially when the public looks for someone to blame. And it no doubt makes for a better movie scene in both "A Night to Remember" and "Titanic."

Bruce Ismay had continued to be active during the sinking, helping women and children to board lifeboats on the starboard side of the Titanic. Eventually, the only boat left on the starboard side was Collapsible C. Ismay helped women and children into it. Depending on the source, he boarded the collapsible or called out for any more women or children, and when none appeared, he boarded the lifeboat.

Collapsible C left the Titanic with about 40 people on board about 20 minutes before the ship sank. It was the last boat to be launched from the davits on the starboard side. Ismay testified to the British Board of Trade that he was at an oar and rowing with his back to the ship so he would not have to see the sinking.

Despite conflicting reports, it is clear that Ismay ultimately survived the sinking, escaping on one of the last lifeboats to leave the ship. This fact has led many to question Ismay's motives and character, with some labeling him a coward who abandoned the ship. However, it is essential to remember that the situation on board the Titanic was chaotic and confusing, and it is difficult to say for sure what any individual's actions truly were.

One of the main points of contention surrounding J. Bruce Ismay's involvement in the Titanic disaster is whether or not he pressured the ship's captain, Edward Smith, to increase the ship's speed to set a new transatlantic record. Some accounts suggest that Ismay did, in fact, encourage Smith to push the ship to its limits, despite the presence of ice in the area. This decision may have contributed to the disaster, as it would have made it more difficult for the crew to maneuver the ship away from the iceberg in time. During the hearings into Titanic's sinking, several references were made to Ismay urging Captain Smith to try for a Tuesday night arrival in New York. There is no credible evidence to support these claims, but the mere   mention of it gave Hearst's yellow journalists more paint to smear a good man's reputation.

This character assassination of Ismay created the image of a man driven by power. "The most serious assertion concerned Ismay's alleged interference with the navigation of the Titanic. Various books and a recent film about the disaster have enlarged upon the Hearst stories."[1] We now see the stereotype of a businessman, the prototype for the J.R. Ewing character in "*Dallas*," only interested in power and money, a man determined to get his way. "If this were the case, one would expect him to have taken every maiden voyage he possibly could, but during his entire business life, and as chairman and managing director of White Star, he took only three; Adriatic in May 1907, Olympic in June 1911 and Titanic in April 1912. Hardly a record for someone in his position, but reflecting the fact that Ismay had more important matters to deal with than worrying about maiden voyages and what the papers might say."[2] White Star Line had 19 ships make their maiden voyages between 1899 and 1912 (the years when Ismay was the Managing Director). He was only on three of them.

Figure 1 J. Bruce Ismay in Washington, DC on his way to the Senate hearing regarding the Titanic's sinking.
Author's collection

It appears that one of the affidavits from the Titanic disaster is based totally on hearsay evidence; the infamous "I heard person A telling person B ...."

Most of this material came to light, not at the American or British inquiries, but at a United States District Court case that began in June 1913 regarding the Oceanic Steam Navigation Company's (the parent company of White Star Line) attempt to limit its liability as owner of the Titanic in the United States. Because of the nature of the action and considering the amount of money involved, witnesses were sought in the hope of providing strong evidence that Ismay had ordered Smith to "make a record," thus, indirectly causing the loss of the Titanic. The evidence provided, at best circumstantial, at worst pure invention, an example of which is Mrs. Emily Ryerson's answers given under cross-examination concerning her conversation with Ismay on board the Titanic:

"Q. He (Ismay) didn't say anything to you about speeding the ship up to get out of the ice?

A. No, that was merely the impression that was left on my mind.

"Q. My question is not whether he spoke about their putting on more boilers and going faster; but I am confining my question to whether he said, or suggested to you, anything that indicated that they were going to increase their speed in order to get out of the ice?

A. As I say, that was merely the impression left on my mind.

"Q. Nothing was said?

A. No, not in so many words — that was the impression left on my mind.

"Q. You don't wish to be understood the Titanic was trying to make a speed record across the Atlantic?

A. I should say my impression was they were going to show — surprise us all by what she could do, on that voyage.

"Q. As a matter of fact, was it discussed whether she should get in on Tuesday night, or Wednesday morning?

A. Yes.

"Q. Among passengers?

A. Yes, and in this conversation with Mr. Ismay also, there was some question about it, because I discussed it with my husband after I got down to the cabin.

"Q. You wouldn't say Mr. Ismay said they were going to make a record?

A. No, I wouldn't say he said those words — his attitude, or his language, we assumed that that was — that we were trying to make a record. I wouldn't say he used those words."[2]

Another thing that makes this testimony hard to swallow, White Star Line had their ships built for comfort, not speed. Because of the Titanic's weight and engine power, there was no way she could match the speed of Cunard's greyhounds, the record-holding Mauretania and her sister, the Lusitania.

Additionally, White Star Line had a habit of slowly breaking in their ship's engines over a few months. If you're old enough to remember buying a new car as late as the early 1960s, you would remember being cautioned not to drive over 40-45 mph for the first 3,000 miles. This was because the state of the art of mechanical engineering at that time wasn't exceptionally fine, and you were doing the final machining on the engine. Similarly, White Star Line's did the same thing with their ships engines.

Then there was Ismay's life after the Titanic. Despite the negative attention he received, Ismay continued to defend his actions in the years that followed. He argued that he had done what he could to help save lives and had not acted recklessly or negligently. However, he also acknowledged that the sinking had been a terrible tragedy, and that he bore some responsibility for what had happened.

The story goes that he became reclusive and withdrawn from society. This is far from the truth. He retired as President of the International Mercantile Marine Corporation on June 30, 1913. Although the media treated this as news (another chance to throw more yellow paint around), Ismay had notified the directors in January 1912, four months before the Titanic sank, of his desire to step down, and it was publicly announced then. Additionally, he remained active as a chairman or director of several large companies, including an insurance company that dealt with claims from the Titanic's sinking, so, no, he could not put the disaster behind him. He and his wife also founded several charities to assist the widows and children of maritime accidents.

So, was J. Bruce Ismay a hero or a villain? As with many historical figures, the answer is not simple. Ismay was a flawed and complex individual who made a number of controversial decisions during his time as managing director of the White Star Line. However, it is also clear that he played a crucial role in the rescue effort on board the Titanic and was deeply affected by the tragedy.

It is also essential to consider the historical context in which Ismay lived and worked. The early 20th century was a time of rapid technological and social change, and the shipping industry was no exception. Ismay was operating in a highly competitive and rapidly evolving industry and was pressured to keep up with the latest advances in shipbuilding and transportation. This may have contributed to his decisions regarding the Titanic and other ships in the White Star Line's fleet.

Ultimately, the story of J. Bruce Ismay and the Titanic disaster is a reminder of the importance of understanding historical figures with nuance and complexity. Ismay was not simply a hero or a villain but a human being with strengths and weaknesses, successes and failures. By examining his life and legacy in detail, we can gain a deeper understanding of the events that led up to the sinking of the Titanic and the complex web of factors that contributed to the tragedy.

As we reflect on the legacy of J. Bruce Ismay and the Titanic disaster, there are a number of important lessons to be learned. First and foremost, we must remember that disasters like this are not the result of the actions of any one individual, but rather a complex interplay of factors that are often beyond our control. We must also recognize the importance of understanding historical figures' complexity rather than reducing them to simplistic caricatures of heroes or villains.

Finally, we must never forget the human cost of tragedies like the sinking of the Titanic. While it is fascinating to delve into the minutiae of Ismay's life and legacy, we must also remember the thousands of individuals who lost their lives on that fateful night. By learning from past mistakes, we can work to prevent similar tragedies in the future and honor the memory of those who perished.

ENDNOTES

[1] Ismay and the Titanic by Paul Louden-Brown, Titanic Historical Society. Excerpt from "The White Star Line: An Illustrated History 1869-1934" Ismay And The Titanic - by Paul Louden-Brown (titanichistoricalsociety.org) Last accessed 4/17/23.

[2] Ismay and the Titanic by Paul Louden-Brown, Titanic Historical Society. Excerpt from "The White Star Line: An Illustrated History 1869-1934" Ismay And The Titanic - by Paul Louden-Brown (titanichistoricalsociety.org) Last accessed 4/17/23.

[3] Ismay and the Titanic by Paul Louden-Brown, Titanic Historical Society. Excerpt from "The White Star Line: An Illustrated History 1869-1934" Ismay And The Titanic - by Paul Louden-Brown (titanichistoricalsociety.org) Last accessed 4/17/23.

CHAPTER 9
Captains Smith, Rostron, and Lord

When people talk about the sinking of the Titanic, the actions that the three captains directly involved took that night almost inevitably come up: Edward John Smith, Arthur Henry Rostron, and Stanley Phillip Lord.

Edward John Smith, Captain of the RMS Titanic. Author's collection.

Edward John Smith was the captain of the Titanic. He went to sea at the age of 17. He received his first command with White Star Line and their ship, the Republic, in 1887 at the age of 37. He failed his first navigation exam but passed on his second attempt the next week, earning his Extra Master's Certificate in February 1888.

Smith was one of the world's most experienced sea captains and the Senior Captain of the White Star Line. As such, he was honored to take each new ship out on its maiden voyage.

When near ice, it was considered standard practice to travel as fast as possible to get through the ice field as quickly as possible. Eleven ship captains testified to that at the British inquiry (see Chapter 10, A Series of Mistakes). But why was this the practice?

I have been unable to find a definitive answer to that question. However, I can give a hypothesis. As the air cools, water vapor condenses and can form fog. Some of the Titanic's personnel reported a slight haze starting to form on the ocean's surface. It would be almost impossible to see an iceberg in a fog, so a ship's speed would have to be reduced to barely making headway. Therefore, get through the ice as fast as possible to avoid a situation where you're stuck and scarcely able to travel.

As a result of his actions, Captain Edward John Smith was guilty of the deaths of 1,496 men, women, and children. Even though he followed the standard practice of the time, he was still negligent in his actions.

Captain Arthur Henry Rostron on board the Carpathia in April 1912. Author's collection.

Captain Arthur Henry Rostron was 15 when he joined the Merchant Navy Cadet School Ship, HMS Conway. In 1894, he passed the exams for his Extra Master's Certificate. He was only 25.

After the Russo-Japanese War, when he served in the Royal Navy, he returned to the Cunard Line and was made First Officer of the RMS Lusitania in 1907. He was transferred to the Bresica the day before the Lusitania's maiden voyage. He became captain of the RMS Carpathia on January 18, 1912.

On the night of April 14, 1912, the Carpathia was four days into its journey from New York to Flume, Austria-Hungary, when her telegrapher, Harold Cottam, heard Titanic's distress call. Rostron ordered Carpathia to turn towards the Titanic's reported position before confirming with Cottam that he was positive it was the Titanic he had heard.

Posting extra lookouts and preparing to take on survivors, Rostron was proactive in preparing for the rescue of the Titanic's survivors. On not hitting any icebergs, Rostron reportedly said, "I can only conclude another hand than mine was on the helm."[1] Without a doubt, Rostron was the hero of the night.

Harold Cottam, circa 1912. Wireless operator on SS Carpathia. Received distress call from the Titanic. Author's collection.

At the same time, he was the most imprudent captain that night. Going full steam into a known ice field where a ship of much newer construction than his had just been sunk, was the height of imprudence. Yet, had he not done it, it's quite likely the number of survivors would have been much less. The ocean had gone from being as smooth as a mill pond to having a slight chop when the Carpathia arrived on the scene.

Captain Stanley Lord of the Californian. Author's collection.

Last is Captain Stanley Lord (no relation to Walter Lord, author of *"A Night to Remember."*) Lord was captain of the SS Californian. He obtained his Extra Master's Certificate at the age of 23. He ordered her to stop earlier in the evening when she encountered a massive ice field, as this was Lord's first encounter with ice at night. Lord ordered his wireless operator, Cyril Evans, to notify all shipping in the area of Californian's position and the ice field. When he got on the air, Jack Phillips immediately rebuked him on the Titanic. Evans, who had already put in a 16-to-18-hour day, shut his radio off and called it a night. Lord would not find out until the following day that Titanic had rebuked the wireless call.

In all fairness, Evans had not prefixed his message with "MSG" (Master's Service Message). By the same token, he had not been told to—yet another example of the young men in the Marconi Company service having more autonomy than they may have been prepared to handle. At the same time, it was a navigational message, and Phillips should not have ignored it. It should have been taken to the bridge immediately, as the Mesaba's message should have been also.

Much has been written and debated about how far apart the Californian and Titanic were on that fateful night. Suffice it to say that when Dr. Robert Ballard discovered the wreck of the Titanic in 1985, it was then proved that the Titanic's calculated position that it was sending out in the distress calls was some 13 miles west of her actual position, based on the wreck's location. By the same token, we don't know if the Californian's position was accurate, as reported that night. The best guess is that the ships were 13-15 miles apart and that had Lord known of the disaster and responded, the

Californian would have arrived at the wreck site about the same time as the Carpathia, given the amount of ice she had to pass through.

Lord and Rostron obtained their Extra Master's Certificates at the ages of 23 and 25, respectively, while Smith was 37 when he received his. Lord was 35, and Rostron was 43 at the time of the accident. Smith was 62. There is evidence of a generational shift in thinking here. Both Lord and Rostron gave more credence to wireless than Smith did. Rostron worked out the new heading for Carpathia to get to the Titanic before questioning his wireless operator, Harold Cottam if he was positive that it was the Titanic he had heard. The younger generation embraced the changing technology that wireless brought, much like the way young people today have adopted to the internet as compared to most people over 75 or so.

"Wireless offered the Titanic the information needed to avoid disaster, but neither company, country, nor captain had developed procedures to ensure that such information was used unequivocally in the best interests of safety."[2]

On pages 3 and 4 of *"Titanic Signals of Disaster,"* the authors devote several paragraphs to listing shipwrecks in the late 19th and early 20th centuries. I've taken the liberty of condensing that information into the following table:

SHIP NAME	YEAR SANK	NUMBER OF FATALITIES
Atlantic	1873	542
Cospatrick	1874	476
Schiller	1875	200
Pacific	1875	236
Duburg	1890	400
Utopia	1891	574
Namehow	1892	400
Norge	1894	≈600
General Slocum*	1904	1021
Suic	1906	350
Republic	1909	6

≈The exact number lost in the sinking of the Norge in 1894 is not known but is estimated at around 600.

*The General Slocum was a paddlewheel excursion boat operating around New York City with day excursions to the north shore of Long Island. It caught fire in the East River, with the

CHAPTER 10
A Series of Mistakes

Was the Titanic destined to sink from the beginning? Did a series of mistakes cause the sinking of the great liner? There is possibly some truth to that theory. Each of the following events contributed to the Titanic meeting the iceberg. If any of these events did not occur, the probability of the Titanic colliding with the iceberg would decrease.

The first event people point to is on September 20, 1911, Titanic's sister ship, the RMS Olympic, collided with the cruiser HMS Hawke as the vessels traveled on parallel courses in the Solent, a straight between the Isle of Wight and England, leading to Southampton. Although Captain Edward Smith was the captain of the Olympic, technically and legally, the ship was under the command of the harbor pilot. The Olympic began a turn to starboard, to which Hawke's captain did not have enough reaction time. The Hawke's bow was designed to allow the cruiser to ram enemy ships. Because of the reinforced bow, two of the Olympics' watertight compartments were punctured. This incident undoubtedly contributed to the idea that the Olympic class vessels were "unsinkable." Emergency repairs were made in

Southampton, taking about two weeks, and then a further six weeks were required for repairs at the Harland and Wolff dry dock in Belfast, Ireland.

Figure 1 A map showing the Isle of Wight and Southampton. The Solent is the inlet that comes from Southampton down to the English Channel. The Isle of Wight divides it into two distinct channels. The Titanic's sister ship, RMS Olympic, collided with the cruiser HMS Hawke in the Solent. Repairs to the Olympic caused the Titanic's maiden voyage to be delayed over one month.

Because of this, the Titanic had to be floated out of the dry dock to allow repairs to the Olympic. This event, which occurred over six months before the Titanic's planned sailing, caused the Titanic's maiden voyage to be delayed by a little more than a month. While that may have impacted the ship's fate, events during the maiden voyage significantly impacted the ship's destiny on the night of April 14-15, 1912.

At noon, the Titanic left her berth in Southampton on April 10, 1912. As the tugboats were guiding the world's largest moveable object at the time, the stern moorings on the SS New York broke, and the stern began to swing out and came within four feet of striking the Titanic's stern. Only quick action by the tugboat captains was able to avoid an accident. However, that incident caused about an hour delay in the crossing.[1,2] Had the SS New York remained firmly moored and the delay not happened, the Titanic would have passed about ¾ mile south of the iceberg[3]. But even had the Titanic missed that particular iceberg, that doesn't mean she would not have collided with another. Remember, Mesaba's warning was of an ice field. I believe her fate was sealed that night; it's just a matter of which iceberg, or icefield, would do the damage.

There were a significant number of Master Service Messages with the MSG prefix that were taken from the Marconi shack to the bridge during the voyage. Many contained warnings of ice in the area the Titanic would be traveling. Despite all those warnings, Captain Smith did not reduce his speed. Was he being imprudent?

At the British Board of Trade inquiry, 11 ship captains testified under oath, as to what they do when in the vicinity of ice. Captain John Pritchard, who formerly commanded Cunard's record-setting Mauretania, which was capable of 26 knots, said "should only slacken speed if the weather conditions were unfavourable." (*The Cambria Daily Leader*, June 24, 1913). At the British Titanic inquiry, he testified under oath that even with "information that there was a probability of your meeting ice on your course" he would maintain speed: "As long as the weather is clear I always go full speed." Pritchard also explained that this was in his experience a "universal practice" - based on his time commanding Cunard ships between Liverpool and New York for 18 years. He also noted that if following the southern track - as did Captain Smith - he had "never got into an ice-field. We do not go North, you know; we go on the southern tracks this time of year." As for lookouts, he would not double them when in "clear weather."[4]

Captain Hugh Young, of the Anchor Line, with 37 years of experience crossing the Atlantic on the New York trade, testified under oath that if ice were reported, he "should keep my course and maintain my speed" in clear weather. He also confirmed this was a "universal practice." [5]

Captain William Stewart, Canadian Pacific, worked for 35 years on the trade between Liverpool and Canada. He was posed with the question if you "were given information that you might meet ice and that your course would take you through the place where you might meet ice, and meet it at night, would you reduce your speed?" His answer was: "No, not as long as it was clear." He was asked further, "If you had information that you might meet field ice, would you still maintain your speed?" and responded similarly: "Until I saw it, and then I should do what I thought proper."[6]

Captain John A. Fairfull of the Allan Line, working the Atlantic for 21 years, also confirmed the evidence of Pritchard, Young and Stewart. He was asked "Is your practice in accordance with theirs?" And he answered "All except that when we get to the ice track in an Allan steamer, besides having a look-out in the crow's-nest, we put a man on the stem head at night."[7]

Captain Andrew Braes, who commanded steamers of the Allan Line for 17 years also confirmed not changing speed or course in good visibility - "Just the same. I never slowed down so long as the weather was clear...I kept my course...I never knew any other practice.".[8]

Captain Frederick Passow, who had been a captain on the North Atlantic for 28 years, in the American Line, and who had crossed about 700 times, testified that he "had a very large experience of ice" and yet did not slacken his speed for ice as long as the weather was quite clear: "Not as long as it was quite clear - no, not until we saw it...when it is absolutely clear we do not slow down for ice.".[9]

Captain Bertram F. Hayes, of the White Star Line, testified that when a position of reported ice he would continue "at the same rate of speed...No alteration...it is the practice all over the world so far as I know - every ship that crosses the Atlantic... Ice does not make any difference to speed in clear weather. You can always see ice then." [10]

Benjamin Steele, marine superintendent at Southampton for the White Star Line, and master mariner with an Extra Master's certificate of 19 years having been at sea "about 26 or 27 years" confirmed the practice of "not slackening speed on account of ice as long as the weather is clear" by responding, "It is. I have never known any other practice."[11]

Captain Richard Jones, master of the SS *Canada* of the Dominion Line, and in the

Canadian service for 27 years testified that his ship was stopped by ice on April 11 1912. However, he also confirmed that after receiving messages about the ice he continued at full speed ahead, considering it a usual practice. He said: "I should think it would be just as safe to go full speed with 22 knots... we always make what speed we can—we always try to get through the ice track as quickly as possible in clear weather."[12]

Captain Edwin Cannons, master with the Atlantic Transport Company with 25 years' experience in the North Atlantic noted that he had "never seen field ice on the southern track." If an iceberg is sighted, he testified that "I keep my speed...Both day and night...I have never had any difficulty to clear when I have met ice ahead." If ice is reported he said: "I should maintain my speed and keep an exceptionally sharp look-out... to maintain speed until the ice is seen." However, if was clear he would not double the look-out.[13]

Captain John Ranson of White Star Line's *Baltic*, on the Liverpool-New York run also confirmed the standard practice: "We go full speed whether there is ice reported or not...We keep up our speed... It has always been my practice." He also stated that it is the practice of all liners on that course, "for the last 21 years to my knowledge." and that he would not double the look-outs at night "not in clear weather."[14]

These 11 captains, with over 248 years of combined experience, present a consistent picture that reports of ice were not a reason for a ship to slow down; therefore, Captain Smith was following the normal procedures of the time. It should be noted that five of the captains were employed by subsidiaries of International Mercantile Marine (IMM) the owners of the White Star Line and the Titanic. That still leaves six captains who were not employed by IMM or its subsidiaries. Captain Smith was following the standards of the time, therefore it's hard to say that he was being imprudent.

The longest-running science fiction program on television, Doctor Who, indirectly addressed what occurred during the Board of Trade hearings; "Answers are easy. It's asking the right questions, which is hard."[15] The hearings were a prime example of the right question not getting asked. The captains were all asked a variation of "What do you do when you are in the vicinity of ice?" And they all answered with a variation of: "As long as the weather is clear, get out of there as fast as we can." The follow-up question was never asked, "Why?"

And that is the key. Why did ship captains go full speed through ice fields to escape the ice?

My research has been unable to find an answer to that question, but I have a theory. As the air cools, the amount of water vapor it can hold decreases. It condenses into clouds and, at ground or sea level, fog. Fog is a grayish-white color, similar in color to an iceberg. Heavy fog would be perfect camouflage for an iceberg. If a ship were caught in heavy fog, it would have to reduce its speed to make up for the reduced visibility caused by the fog. At the same time, the slower speed means less water is flowing over the rudder, meaning the ship will be less responsive to steering commands. I'm more than willing to consider another explanation, but this is the only explanation I've been able to come up with that makes sense.

As previously stated, there was one ice report that did not make it to the bridge. The report from the SS Mesaba was received on the afternoon of April 14; however, it was not prefixed with MSG but rather SG and ended up getting lost in the shuffle of messages that passengers were sending on the Titanic. Can Marconi Operators Phillips and Bride be faulted for not taking that message to the bridge?

The wireless operators existed in a no man's land on the ship. While they were assigned to the ship, they were not employees or officers of the ship; instead, they were employees of the Marconi Company. At the same time, the shipping companies considered the Marconi operators equivalent to their Junior Officers. By not being ship officers, they could not access navigational data. Even if they did have access to the Titanic's course and speed, there would be the question as to whether or not they would have known that the ice reported by the Mesaba was directly in the Titanic's path.

Also, radio was not consistently considered a navigational aid at that time. The Marconi Company's main business was sending messages from the ship's passengers to friends or business associates on shore. Master Service Messages (MSG or SG prefixes) were a courtesy for the vessels. At one of the hearings, it was established that all navigational messages would be treated as priority messages, whether or not they had the MSG prefix. The ship captains did not view the Marconi system as an efficient means of communication. There's a photograph of the Californian, taken from the Carpathia when both ships were at the Titanic wreck site, and it's evident that the two vessels are communicating via signal pennants even though both are equipped with radios. We know they are communicating via pennants because the Californian has raised pennants, and the first pennant is "Answer."

Figure _ the SS Californian as seen from the deck of the Carpathia on the morning of April 15, 1912. This image has been rephotographed so many times that, except for the pennant which indicates "Answer" the rest of the message is unreadable. Author's collection.

As previously mentioned, at 11:00 p.m. AST on April 13, the Titanic's primary Marconi transmitter broke down. The standard procedure in such an event would be to use the ship's auxiliary transmitter and allow a Marconi engineer to repair the equipment while the ship was in port. The auxiliary transmitter on the Titanic was estimated to have a range of only 50 miles. Also, at the time, wireless operators were not allowed to communicate with wireless operators from competitor companies. Additionally, the Marconi Company (and the other wireless companies) made their money by sending messages for the passengers. A typical 10-word Marconigram would cost 8s 4d (8 shillings, 4 pence, or about one hundred pounds in today's money, adjusted for inflation). The Marconi company required that their telegraphers be able to send at 25 words per minute; they could theoretically send one 10-word message each minute (including the additional address and ancillary message data) for revenue of 8s 4d per minute. That's about 45 messages an hour or 1,080 messages per day (allowing for the human factor of less than 100% efficiency). Yes, the Marconi wireless was a novelty for first-class passengers but an extremely profitable novelty for the Marconi company.

Given those operating restrictions and the tremendous backlog of passenger messages (read: money), Phillips and Bride decided to break the rules and attempt to repair the primary transmitter themselves. Most of the repair took place at a time when Phillips should have been sleeping. That lack of sleep would manifest itself the next day when, in reply to an ice warning from the Californian,

Phillips angrily sends back, "Shut up! Shut up! I'm working Cape Race."

Had Phillips and Bride not repaired the transmitter, what would the likely outcome have been? At the time of the accident, the RMS Carpathia was approximately 50 miles from the Titanic, and there was some question as to whether or not the two ships would have been able to communicate. The decision to attempt to repair the primary transmitter and its success undoubtedly contributed to the rescue of the Titanic passengers. On the other hand, had the repair not been made, would Phillips have been working Cape Race via a relay with the Californian? Probably not. Cyril Evans didn't hear Cape Race when he started to broadcast the ice warning, and he had a long day; he may have shut down his radio at 11:30 anyhow. If that were the case, it's entirely possible that no one would have heard the Titanic's distress call, and everyone likely would have perished. The Carpathia encountered stormy weather returning to New York, and that same weather would likely have affected the Titanic's lifeboats. At the time of the accident, it was not required that there be a 24-hour watch on the Marconi wireless. That would come about as one of the recommendations from the British Titanic Inquiry.

There was one other mistake made after the collision with the iceberg. Captain Smith ordered the distress rockets to be fired at five-minute intervals. The regulations at the time required distress rockets to be fired at one-minute intervals.

ENDNOTES

[1] On A Sea of Glass, Third Edition, 2015, Amberly Publishing, pp 80-82

[2] www.encyclopediatitanica.org/the-new-york-incident Last accessed August 20, 2022.

[3] Author's calculations based on the speed of the Labrador Current in the area where the collision occurred and had the Titanic arrived at that location one hour earlier (assuming the SS New York had not broken her moorings).

[4] British Titanic Inquiry, Day 27

[5] British Titanic Inquiry, Day 27

[6] British Titanic Inquiry, Day 27

[7] British Titanic Inquiry, Day 27

[8] British Titanic Inquiry, Day 27

[9] British Titanic Inquiry, Day 21

[10] British Titanic Inquiry, Day 21

[11] British Titanic Inquiry, Day 21

[12] British Titanic Inquiry, Day 24

[13] British Titanic Inquiry, Day 24

[14] British Titanic Inquiry, Day 26

[15] https://en.wikiquote.org/wiki/The_Doctor
Doctor Who, The Face of Evil, January 1977

The American and British Titanic Inquiry transcripts are available at www.titanicinquiry.org

Figure 4 Titanic's surviving officers, left to right Fifth Officer Harold Lowe, Second Officer Charles Lightoller, Fourth Officer Joseph Boxhall, seated is Third Officer Herbert Pittman. The Titanic would remain a dark stain on their maritime careers and none of them would receive a merchant command of their own. Photo is public domain.

Second Officer Charles Lightoller, the highest-ranking officer to survive the Titanic, would never receive his own command. Neither would the other three senior surviving officers; Boxhall, Pittman, and Lowe. The Titanic would haunt their maritime careers for the rest of their lives.

As referenced earlier, Lightoller quotes a conversation he supposedly had with Jack Philips, the senior Marconi operator, while standing on the overturned Collapsible B. That conversation was never mentioned in any of Lightoller's testimony, although it would appear to be highly significant to the events of the night of the sinking.[3] Why would he not bring that up during either of the hearings? Also, this conversation's validity is possibly why his book was withdrawn from publication when the Marconi Company threatened legal action.

You will recall that ATS on April 14 was 2 hours 58 minutes behind GMT and 2 hours 2 minutes ahead of New York. And Fourth Officer Boxhall stated that 11:46 p.m., when he calculated the distress coordinates, equated to 10:13 p.m. in New York, a difference of 1 hour 33 minutes ahead of New York, not the 2 hours 2 minutes calculated at noon.

Lightoller would testify that 2:20 a.m., the time of the sinking, would equate to 5:47 a.m. GMT, a difference of 3 hours 27 minutes. One hour 33 minutes plus 3 hours 27 minutes equals 5 hours, the time difference between New York and GMT.

Lightoller, Boxhall, Third Officer Herbert Pitman, and Fifth Officer Harold Lowe were the only senior officers of the Titanic to survive the sinking. They spent time together on the Carpathia on the return trip to New York. There was ample time to agree on their stories; would they have colluded? None of the officers gained anything; none received commands of their own. The Titanic disaster was a black mark on their service records.

My opinion is that, while there was no collusion in the strict legal sense of wanting to defraud an organization, there was, however, a "meeting of the minds," an understood "contract" to make sure their testimony would add up correctly.

I have mixed views on this. There is a report from Junior Wireless Operator Bride that he saw Philips dead on Collapsible B. Yet there is also evidence that Philips was running toward the stern as the ship sank.

This creates a problem. Suppose Philips never made it to Collapsible B. In that case, the conversation that Lightoller says he had with Philips in Titanic and Other Ships is a complete fabrication, and it also explains why Lightoller did not mention it at either hearing--because there would have been the other survivors from Collapsible B who could contradict him. And, if Lightoller was willing to lie about this, how much else did he lie about during the inquiries? This raises another question. Who did Bride see on Collapsible B that he thought was Jack Philips?

Let's look at the other side and, for the sake of argument, assume that Philips did make it to Collapsible B and that Bride did see Philips' body on B. That doesn't change the situation that Lightoller's conversation with Philips was a complete, or mostly complete, fabrication. And again, it explains why there was no mention of it in either inquiry because other survivors could contradict him. I can see both sides; unfortunately, neither does Lightoller's reputation any good.

ENDNOTES

[1] Titanic and Other Ships, Charles Herbert Lightoller. Chapter 35. Project Gutenberg of Australia ebook. Most recent update April 2005. Last accessed October 19, 2022.

[2] Titanic and Other Ships, Charles Herbert Lightoller. Chapter 35. Project Gutenberg of Australia ebook. Most recent update April 2005. Last accessed October 19, 2022.

[3] Titanic and Other Ships, Charles Herbert Lightoller. Chapter 35. Project Gutenberg of Australia ebook. Most recent update April 2005. Last accessed October 19, 2022.

Chapter 12
THE MARCONI COMPANY PRACTICES

Did the Marconi Company engage in proper practices when training their operators? Most of the operators were between 20 and 25 years of age. Were they adequately equipped to handle the responsibilities that came with their position?

Marconi operators initially received six weeks of training, although later, it would expand to be a multi-year correspondence course. Was that enough training? Relatively speaking, it was a good-paying job,1 pound sterling (about $5) per week.

Jack Phillips was one of the earlier graduates of the Marconi school. In theory, he should have known that all navigational messages would be treated as "urgent" and taken to the bridge when received. Yet he failed to do this on the night of April 14, 1912.

The Mesaba's Ice Warning warned of an ice field directly in the path of the Titanic. Based on the wreck's location, the Titanic struck an iceberg in the ice field reported by the Mesaba. The information for this Marconigram came from page 29 of Titanic

Signals of Disaster by John Booth and Sean Coughlan.

According to the Marconi Company, all the Marconigram messages submitted to the British Board of Trade inquiry had been stamped "COPY," and this one has that stamp.

Wireless was a new technology. The earliest use I can find of it was in 1899 with the message "Sherman is sighted," the wireless message said, referring to the troopship Sherman, which was returning a San Francisco regiment from the battlefields of the Spanish-American War. It marked the first use outside England of this technology that was still in its infancy. It was sent by Lightship Number 70 to a receiving station at the Cliff House in San Francisco.[1] The article references that prior use of wireless on ships had occurred in England, but no date is given.

New technology. There's the key. There are multiple references to a photo of the Carpathia and Californian at the Titanic wreck site. Both ships were equipped with wireless, yet they were using traditional signal flags to communicate.

Figure 1 Copy of Mesaba's Ice Warning included in the set of Titanic related messages for the British Board of Trade inquiry.

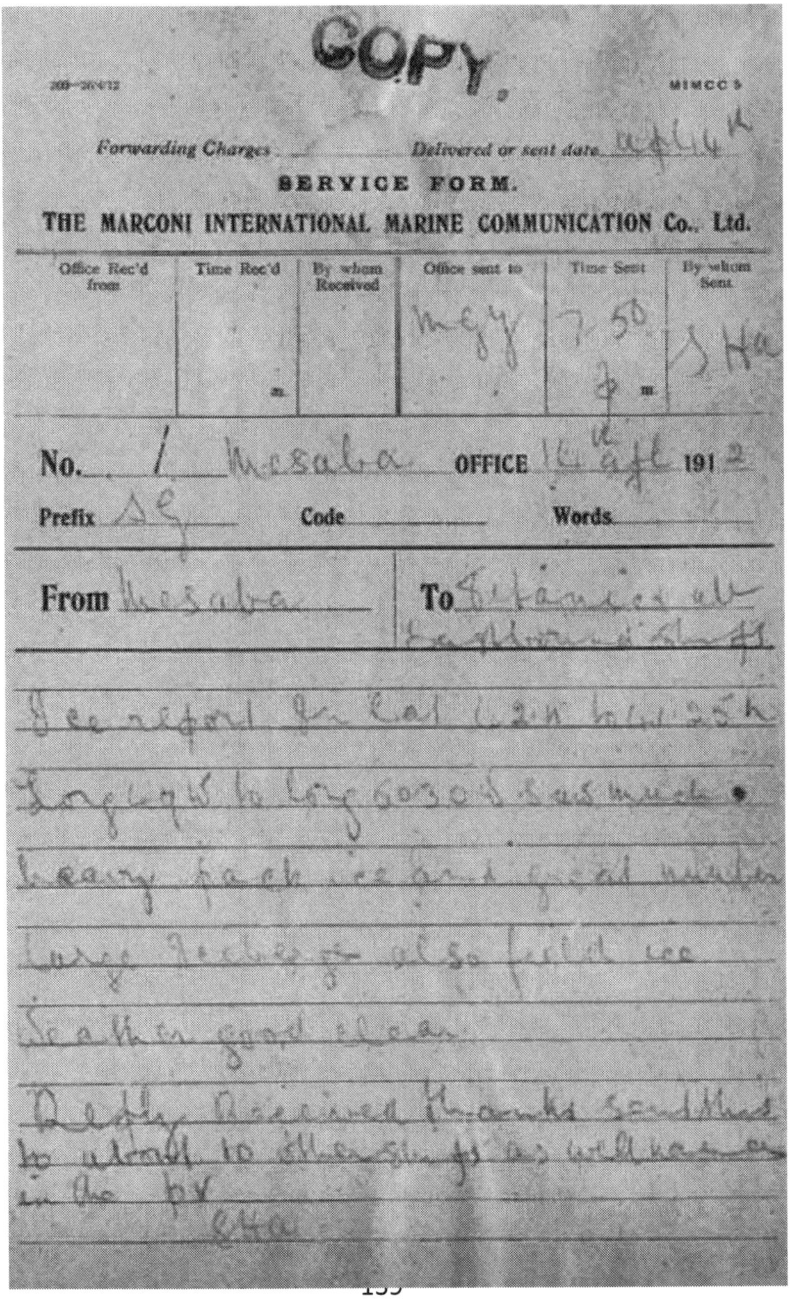

COPY.

209—20/4/12 MIMCC 5

Forwarding Charges — *Delivered or sent date* Apl 14th

SERVICE FORM.

THE MARCONI INTERNATIONAL MARINE COMMUNICATION Co., Ltd.

Office Rec'd from	Time Rec'd	By whom Received	Office sent to	Time Sent	By whom Sent
	m.		MGY	7.50 p. m.	SHa

No. 1 Mesaba **OFFICE** 14th Apl **191**2

Prefix SG **Code** **Words**

From Mesaba **To** Titanic and all Eastbound ships

Ice report in lat 42 N to 41.25 N

Long 49 W to long 50.30 W saw much

heavy pack ice and great number

large icebergs also field ice

Weather good clear

Reply Received thanks send this

to all other ships as well as on

in the PV

SHa

Californian is flying communicates "Answer." Figure 2 is apparently the image that has been referenced, there are signal flags on the Californian, but the resolution of the image, makes it difficult to determine the message being conveyed by the flags because of having been rephotographed through multiple generations; signal flags are read from the top down, and the first flag (or pennant) the Californian is flying is the "Answer" pennant.

Figure 2 the SS Californian as seen from the deck of the Carpathia on the morning of April 15, 1912. This image has been rephotographed so many times that, except for the pennant which indicates "Answer" the rest of the message is unreadable. Author's collection. The "Answer" pennant also has a second meaning, depending on the context. That is, "Message received and understood."

Another reason why the ships are communicating via signal flags can be their relative proximity. As there was no frequency control on the radios at that time, why tie up a frequency for several hundred miles when you could use flags to communicate the several hundred yards?

At the same time, the Marconi Company operators did not necessarily have the judgment to do their jobs adequately. The Titanic received eight ice reports on April 14th. The Mesaba ice warning was not passed to the bridge. It was obviously a navigational message (see Figure 1), but Jack Phillips chose not to send it to the bridge. I can almost understand his reasoning. We've already taken six reports; what difference will one more make? It may not have made a difference. Then again, it might have saved 1496 lives and the ship. Jack Phillips did his job, but not totally. I'm not saying that he was negligent. I am saying that he was lacking good judgment.

The Titanic tragedy proved the usefulness of having wireless on ships. Before wireless, surviving the accident would be purely a matter of chance if a ship were in an accident. Theoretically, enough ships were plying the oceans that rescue should occur in 24-48 hours. The practical side was quite different. No one outside of the survivors would know about the accident, and searchers would have no idea where to look for the survivors. By the time a ship would be reported as overdue, the odds of there being any survivors were between slim and none, and slim was leaving the building. And the survivor's lot wasn't much better. Depending on where the accident occurred, survivors could look at a few days to maybe a week of survival before dying of exposure, starvation, thirst, or a combination of all those causes. Even though Jack Philips is responsible for not taking the Mesaba's message to the bridge, once the accident happened, he did everything in his power to get help to the Titanic as quickly as was humanly possible.

Both of Titanic's wireless operators had attended the Marconi College in Liverpool; chief operator Jack Phillips in 1906 and junior operator Harold Bride in 1911.

"By the standards of the new science of wireless, Jack Phillips was an experienced operator, having attended the Liverpool college and begun operating shipboard wireless only six years after wireless equipment had been installed on a merchant vessel."[2] Bride was, by comparison, inexperienced, having been on ships only nine months. He was not lacking in confidence, as his subsequent work assisting Carpathia's wireless operator, Harold Cottam, showed.

The wireless operators were in a kind of "no man's land" when it came to being on a ship. "In terms of the day-to-day running of the ship, this left wireless as an additional (and perhaps optional) form of assistance, rather than an integral and vital source of information."[3] Titanic's captain, E.J. Smith, had spent virtually all of his life in command of vessels without the benefit of wireless. Before the Titanic sank, you have to wonder how vital wireless was to seamen of his generation. Did they understand the potential benefits it offered? Or did they think of it as more of a novelty or gimmick?

The other side of the proverbial coin was that the Marconi Company made money from the messages it sent for paying customers, i.e., the first-class passengers. Navigational messages, ice warnings, etc., were handled as a courtesy for the shipping companies; they were not money-makers. A wealthy passenger asking that his private railcar be waiting in New York on Wednesday morning for him or a passenger sending an order for a stock sale or purchase would be a money maker for the Marconi Company; a message warning of ice somewhere ahead of the ship would not make money.

Although there was a priority sequence for messages, e.g., navigational messages were to be given top priority, which in theory would mean they would not need the "MSG" or "SG" prefix to be taken to the bridge, in practice, the wireless operator had a great deal of autonomy in handling the messages.

"Phillips, who went down with the ship, might be forgiven for believing that sufficient information about ice had already been passed on to the ship's officers. Eight ice warnings had so far been received and taken to the bridge that day, and three had been acknowledged by wireless."[4] Logically, they could assume that the captain was well aware of the danger to the ship. Ice in the North Atlantic shipping routes was not uncommon in April, and ice warnings, thanks to wireless, were just as abundant.

Despite the plethora of warnings, the Titanic sped on. It has been theorized that the Titanic was out to capture the Blue Riband for the fastest crossing of the Atlantic. That theory, pardon the pun, doesn't hold water because the Titanic was so much heavier than the Cunard liners Mauretania (the holder of the Blue Riband from 1907 until 1929) and Lusitania that trying to outrun them would be impossible. Perhaps Titanic was trying to beat the crossing time of her sister, the Olympic. That could be. Or was Captain Edward Smith so confident from his over 40 years at sea that he felt it unnecessary to give the wireless messages the consideration they deserved? We will never know the answer to that question. It went down with E.J. Smith.

What has survived the century-plus since the disaster is Phillips' "sense of urgency in sending commercial messages becoming more apparent as the night went on."[5] When the SS Californian stopped for the night because she was surrounded by ice, her captain, Stanley Lord, ordered his wireless operator, Cyril Evans, to send the word out to nearby ships a strong rebuke from the Titanic. As mentioned previously, Captain Lord would not find out about the rebuke until the next morning.

ENDNOTES

[1] Wired Magazine website, August 23, 2011, <u>Aug. 23, 1899: First Ship-to-Shore Signal to a U.S. Station | WIRED</u>. Last accessed 3/15/2023.

[2] Titanic Signals of Disaster, 1993, White Star Publications, p. 24

[3] Titanic Signals of Disaster, 1993, White Star Publications, p. 24

[4] Titanic Signals of Disaster, 1993, White Star Publications, p.30

[5] Titanic Signals of Disaster, 1993, White Star Publications, p.31

CHAPTER 13
What if...

Since the sinking of the Titanic, many people have speculated about what might have happened if the ship had not hit the iceberg. Astor, Guggenheim, and Straus were reportedly opposed to the creation of the Federal Reserve Bank, favored by J.P. Morgan, the owner of International Mercantile Marine (which included the White Star Line and, therefore, the Titanic). According to multiple online conspiracy theories, the Titanic and Olympic were "swapped," it's the Olympic at the bottom of the Atlantic Ocean. The theory goes that Captain Smith had orders to scuttle the ship to get rid of Astor, Guggenheim, and Straus. Olympic's hull number was 400; the Titanic's was 401. All photos show hull number 401 sitting some two and a half miles below the Atlantic.

The fascinating thing about all the "what if" scenarios is they all assume the Titanic makes it safely to New York. They don't consider the full texts of the Baltic's and Mesaba's ice warnings. Both sets of warnings accurately pinpointed the iceberg that the Titanic struck, taking into account the iceberg's drift from the time of the sighting. The Mesaba's warning also indicated the presence of a massive ice field about three miles west of the iceberg.

Figure 1 William McMaster Murdoch, Titanic's First Officer and Officer Of the Watch (OOW) when the collision occurred. Author's collection.

We know the ice field was there. It had stopped both the Mount Temple and the Californian. The Californian tried to warn the Titanic about this field when Phillips rebuked their call, and that message, even though it was a navigational message, never made it to Titanic's bridge.

Now that we have an accurate picture of what was lurking three miles or seven to eight minutes cruising time beyond the iceberg, we can engage in a more precise "what if" scenario.

Keep in mind that steam engines can't be immediately reversed. Yes, the officer of the watch can send the message to reverse the engines immediately by the ship's telegraph, but it will still take time for the engines to stop because of the inertia of the mechanism. Assuming the engine is making 70 rpm, it could take 45 seconds to one minute, if not longer, to come to a stop. Only then can the engines be reversed. The ship weighs just over 52,000 tons. With that much weight traveling forward at 21 knots, it's going to take quite a while to come to a stop. The engines will be reversed while the ship is still traveling forward. Cavitation will most likely occur around the propellers, meaning that the propellers will create a vacuum as they try to stop the ship, resulting in a considerable loss of efficiency.

Let us assume the Titanic misses the iceberg by a quarter-mile. It doesn't matter whether it misses to port or starboard; it's still missed. Or, for this discussion, you can even assume the iceberg wasn't there.

The Titanic is still cruising along at 21 knots. About eight minutes later, Titanic comes upon the ice field stretching as far as the eye can see both north and south. Is a collision inevitable? Put yourself in William McMaster Murdoch's shoes. He was the Titanic's First Officer and Officer of the Watch (OOW) at the time of the collision. What do you do?

We can extrapolate some data by reviewing the Californian's actions that night. A white haze appeared on the horizon, and about three minutes passed before it was recognized as part of a massive ice field stretching from the northwest to the southeast as far as the eye could see. Without getting into the physics, the ship was turned in a starboard direction (hard a port in the helm commands of the day) and avoided the ice.

The Californian was traveling at about 11 knots, roughly half the speed of the Titanic. If the Titanic passed the iceberg, her crew would see the field ice at about the same distance the Californian's crew did or after about 1½ minutes of cruising. Murdoch would probably order hard a port, and the bow would turn to starboard. Again, the Titanic would have barely avoided disaster.

But what happens if Murdoch doesn't recognize the ice for what it is until it's too late? Does he opt for a head-on collision or a broadside collision? Those are, fundamentally, the two choices that Murdoch has. Whichever choice he makes will result in many injuries and deaths.

A head-on collision would likely collapse the first two watertight compartments. There would be many firemen killed, as well as Third Class male passengers traveling alone. How many more hull plates are damaged; how many watertight compartments are compromised? It's possible the force of the impact could have warped some of the watertight compartment door frames, destroying their integrity.

I'll admit that I don't have the Titanic's construction specifications or the knowledge of engineering or physics to calculate what damage might occur in a head-on collision. My gut tells me the ship would probably sink, and no one would know where. If the third watertight compartment collapses, the foremast will collapse, bringing down the Marconi antenna and probably shorting out the entire wireless system in the worst case, as the antenna would be electrically grounded to the ship's hull. Even if any of the lifeboats got away, the odds of finding survivors would be mighty slim because no one would know to look for them. Remember, the wireless had already failed once. Probably for at least 12 hours, there would be the assumption that the wireless had failed again. Only then, when she hadn't been heard from, would it possibly occur to people that the Titanic had been in an accident—and no one would know where to look.

Let's assume the ship retained its watertight integrity. Her forward speed would be significantly reduced, as many of her firemen would have been killed or injured in the collision. The wireless might be inoperative if the foremast had collapsed, dropping the wireless antenna to the hull. How many passengers and crew throughout the ship would have been injured by the sudden stop? Probably passengers who were only slightly injured, or not injured at all, would have been asked to help with aid and comfort to the more seriously injured. I would guess that somewhere between 50 and 70% of the passengers and crew would have been injured or killed.

A sideways collision is much easier to reason out. Probably between six and ten of the watertight compartments would be compromised. With the water coming in on one side of the vessel, she would list towards that side. The davits on that side would be useless because of the proximity to the ice, and the lifeboats would be swinging away from the ship, so far away that passengers couldn't jump across. The davits on the other side would be useless because the hull would rapidly approach the point where capsizing was inevitable. Lifeboats would be lowered onto the hull. This is the worst-case scenario. No one survived the accident, and over 2200 people died as she capsized probably in under an hour. Although under this scenario, the world would possibly know what happened because the wireless would likely have remained operational for a time.

CHAPTER 14

Conclusions

What can we conclude from the tragedy of the Titanic's sinking?

Jack Phillips and Harold Bride were competent wireless operators. Despite that, they were lacking in judgment. On April 14th, they had taken six ice warnings to the bridge—the warning from Mesaba told of ice directly in the Titanic's path. Phillips acknowledged the receipt of the warning, but it was never taken to the bridge. Harold Bride was virtually useless as a witness. His account of what occurred changed with virtually every telling. The only consistent facts were that he was a wireless operator on the Titanic, it hit an iceberg, and the ship sank. An entire book could probably be written about Bride's continually changing testimony.

Although the Marconi Company said that all navigational messages would be taken to the ship's bridge immediately, this did not happen on the Titanic. Remember that all navigational messages were handled as a courtesy for the ships; the money was in messages to and from passengers. Are we really surprised that Phillips and Bride would violate Marconi's policy and repair the primary wireless unit instead of using the backup transmitter and having the primary fixed in New York? Since they violated that policy, it's no great leap of logic to say they would violate the policy and ignore the most critical ice warning when they had already taken six warnings to the bridge that day.

Would the bridge receiving the ice warning from the Mesaba have made a difference? Actions speak louder than words. Captain E. J. Smith was 62. He apparently did not have the appreciation for wireless as a navigational aid that the younger captains, Lord and Rostron, had that night. Smith's actions appeared to be those of a mariner who feared nothing. Despite the multiple ice warnings, Smith continued to drive the Titanic at 21½ knots, close to her maximum speed. Even though he knew there was a strong possibility of Titanic encountering ice, he took no additional actions to protect his ship.

As we know, RMS Titanic only carried 20 lifeboats, with room for roughly half the approximately 2200 passengers and crew on her maiden voyage. The British Board of Trade used their hopelessly outdated regulations based on the ship's gross tonnage, not the number of passengers she could carry. To blame any other individual or organization other than the BOT for the lack of lifeboats is the height of arrogance. Even as the number of lifeboats for a ship the size of Titanic was being approved, other countries realized that basing the number of lifeboats on the ship's tonnage instead of the number of passengers she could carry was looking at things the wrong way.

Second Officer Charles Lightoller's story about the events of that night remained remarkably consistent—until he wrote *"Titanic and Other Ships."* With the publication of that book, his story about the sinking begins to lose credibility. He makes statements that, if true, should have been made at the U.S. Senate inquiry or the British Board of Trade inquiry. The fact that they weren't made at either inquiry, coupled with the fact that the publisher removed the book from store shelves after Mr. Marconi threatened to sue, casts a long shadow of doubt on the testimony of Charles Herbert Lightoller.

Beyond any doubt, the most misunderstood man that night was J. Bruce Ismay, the managing director of the White Star Line. Multiple witnesses stated that he helped many women and children into the lifeboats that night. In the end, he did get into a lifeboat. He did not take someone else's place. Had he not boarded the lifeboat, one more fatality and one less person would have been saved. Mr. Ismay was unfortunate enough to have William Randolph Hearst as an enemy, and the result of surviving the sinking of the Titanic was his reputation being ruined.

Titanic would provide a lesson in hubris. She was the largest moveable object in the world at the time of her launching, and she was, without a doubt, the most luxurious ship in the world. Even third-class—steerage passengers—had it better than most other ships. There were ships engaged in the transatlantic trade that would carry third-class passengers from Europe to America, then use the third-class space to bring cattle from America to Europe on the return trip.

Who is to blame for the sinking of the Titanic? There was plenty of blame to go around that night. Captain Edward John Smith was ultimately responsible for the accident and subsequent sinking of the Titanic, and, as captain, he bears ultimate responsibility for the accident. But he's not the only one to shoulder the blame. The Marconi operators, John George "Jack" Phillips and Harold Sydney Bride, and ultimately, the Marconi Company have some responsibility. We don't know why Jack Phillips ignored the established policy and did not promptly take the warning from the Mesaba, and other navigational warnings, to the bridge.

Whether it was an error in Phillips' judgment or due to a lack of training at the Marconi wireless school, he still did not follow the proper procedures. We know that he did not always follow the rules; repairing the primary transmitter is evidence of that. Granted, after the collision, he did all he could and succeeded in getting help to come, but that does not absolve him of his failure to get the Mesaba's ice warning to the bridge promptly. There is no guarantee that had the warning reached the bridge in a timely manner, there would have been a different outcome, but there was a communication breakdown.

I have deliberately omitted the "mystery ship" seen from both the Titanic and the Californian. On both ships, the officers estimated the distance to the "mystery ship" at three to five miles. The British Board of Trade concluded that the "mystery ship" seen by the Californian was the Titanic and vice versa. The Board of Trade concluded that the Titanic's position was correct, while the Californian's was in error. Thanks to Dr. Robert Ballard's discovery of the wreck, we now know that the Titanic was 13 miles east of where she was reporting herself to be that night. The conditions that night favored the formation of a cold weather superior mirage, but we don't know that one formed. Just because the conditions were favorable for creating a mirage doesn't mean it happened. It's like favorable conditions for a hurricane to develop, but it doesn't happen. I don't know whether or not there was a "mystery ship." My belief is that the "mystery ship" more than likely did not exist.

I believe that the Titanic and Californian were between 18 and 20 miles apart, which would put them below the horizon as seen from the other ship. I believe that a cold weather mirage (see Chapter 3) occurred, which allowed the ships to see each other via atmospheric refraction. The ships would have appeared to be above the horizon, but with it being a dark, moonless night, it would be impossible to tell that they were above the horizon.

Titanic's socket signals were designed to explode at an altitude of 600 to 800 feet, and would produce a shower of stars. They would also be visible for a range of about 30 miles. The distress signals were supposed to be fired at one-minute intervals. Fourth Officer Boxhall was firing them at five-minute intervals.

At some point in the mirage, the column of air above the ships would equalize in temperature. This would result in the mirage ending. Suppose the socket signals were exploding at an altitude that appeared to be maybe 100-150 feet above the image of the Titanic. That would account for the "queer" appearance of the mystery ship as seen from the Californian.

There is, of course, another possibility. That is that there was, indeed, a mystery ship very close to the Titanic that night. If that's true, then there is a guilty party who could probably have rescued additional people from the sinking. Unfortunately, we'll likely never know because you can just about bet the farm that their scrap log either went overboard or ended up as fuel in one of their boilers.

I've presented the evidence, so who do YOU think is responsible?

Bibliography

U.S. Senate Inquiry into the sinking of the R.M.S. Titanic. Accessed from www.titnaicinquiry.org . Last accessed December 14, 2021.

British Board of Trade Inquiry into the sinking of the R.M.S. Titanic. Accessed from www.titanicinquiry.com . Last accessed December 14, 2021.

Lord, Walter, A Night To Remember. Copyright 1955, 1983, Walter Lord. St. Martins Press, 175 Fifth Avenue, New York, NY 10010.

Fitch, Tad, Layton, Ken, et al., On A Sea of Glass. Third Edition, Copyright 2015 Tad Fitch, J. Kent Layton, Bill Wormstedt. Amberly Publishing, The Hill, Stroud, Gloucestershire, UK, GL5 4EP

www.encyclopedia-titanica.org

Hughes, Michael, Bosworth, Katherine, Titanic Calling Wireless Communications During the Great Disaster. Copyright 2012 Bodleian Library, University of Oxford, Great Wall Printing Ltd., Hong Kong

Booth, John and Coughlan, Sean, Titanic Signals of Disaster, Copyright 1993 White Star

Publications, 30 Edenvale Road, Westbury, Wiltshire, UK

Matthews, Rupert, <u>Titanic the Tragic Story of the Ill-Fated Ocean Liner.</u> Copyright 2022 Arcturus Publishing Limited, 26/27 Bickels Yard, 151-153 Bermondsey Street, London, SE1 3HA England

Lightoller, Charles H, <u>Titanic and Other Ships</u>, Copyright 1935 Online version accessed from Project Gutenberg Australia

Gracie, Archibald, Colonel, <u>Titanic: A Survivor's Story</u> (originally published as <u>The Truth About the Titanic</u>) Copyright 2008 Project Gutenberg Press, US

Made in the USA
Columbia, SC
07 March 2024